2002

GUIDING ICARUS

GUIDING ICARUS

Merging Bioethics with Corporate Interests

RAHUL K. DHANDA

Interleukin Genetics, Inc.

Foreword by

PHILIP R. REILLY

Interleukin Genetics, Inc.

 **WILEY-LISS**

A JOHN WILEY & SONS, INC., PUBLICATION

Copyright © 2002 by John Wiley & Sons, Inc., New York. All rights reserved.

Published simultaneously in Canada.

For ordering and customer service, call 1-800-CALL-WILEY

Library of Congress Cataloging-in-Publication Data:

Dhanda, Rahul K.
   Guiding Icarus : merging bioethics with corporate interests / Rahul K. Dhanda ; foreword by Philip R. Reilly.
      p. cm.
      ISBN 0-471-22380-8
   1. Biotechnology—Moral and ethical aspects.   2. Biotechnology industries.   I. Title.
   TP248 .23   .D48 2002
   174'.96606—dc21   2001006726

Printed in the United States of America

10 9 8 7 6 5 4 3 2 1

To my parents

Krishan K. Dhanda
for the will, compassion, and humanity that survives him

and

Shashi Dhanda
for her strength, love, and courage

CONTENTS

FOREWORD

In the spring of 1945, as they rushed to build the atomic bomb, American physicists initiated what was to become perhaps the first global debate about the use of a new technology. With just a bit of imagination, one can reconstruct the outlines of hushed conversations that unfolded during late night walks along the dirt roads within the hastily constructed secret installation in Los Alamos, New Mexico. Some of the nuclear wizards, epitomized by Edward Teller, were resolute in their belief that the construction of atomic weapons was essential to winning World War II and, later, containing Communism. Others, like Hans Bethe, the German immigrant who had worked out the equations that define the nuclear fires of stars, were much more dubious. During the late 1940s, he and many of his colleagues would warn the world about the dangers inherent in the new tools they had forged. The awesome power of nuclear technology was crystallized in the words of J. Robert Oppenheimer, the Princeton physicist who led the Manhattan Project. Witnessing the explosion of the world's first nuclear device in Alamagordo, New Mexico, he reflected, "Now, we are all destroyers."

If he meant that the legacy of the Manhattan Project would be the capacity for self-destruction, Oppenheimer was right. Fortunately, despite the vast environmental damage wrought by atmospheric testing of mammoth hydrogen bombs in the 1950s, the charter members of the nuclear club were able (just barely, at times, it seemed) to control their awesomely powerful tools. For 45 years, the Cold War was waged with the horrible threat of nuclear annihilation lurking in the shadows of the corridors of power. In 2002, it is no small irony that the first truly dramatic reduction in the number of nuclear warheads held by Russia and the United States grew out of the horrific success of a small group of antitechnology terrorists whose leader lived in the caves of Afghanistan.

The Atomic Age, as it was called in the 1950s, was arguably the first of a series of transforming events arising out of mankind's ever deepening mastery of nature. Since then, the pace of technological transformation has increased steadily. The changes brought about by the rise of computing and the Internet (the origins of which can be traced to the arms race that created the nuclear arsenals) in the 1980s and 1990s—developments which make "globalization," the socioeconomic trend most disturbing to fundamentalists of any stripe, inevitable—will be, many think, even more profound than those that flowed from our ability to control thermonuclear physics.

Developments wrought through biotechnology—which I use here to mean the application of a set of biochemical tools to study and manipulate DNA, RNA, and proteins—will, I firmly believe, have an impact on humanity that eclipses these other major technological transformations. For unlike the peaceful use of nuclear energy and computing, the quintessential goal of biotechnology is, in the deepest sense, to change nature itself. Although we are in the earliest moments of learning how to use and improve our powerful tools, the hazy outlines of the future are already discernible. In time, humans will use biotechnology to redefine our relationship with the world. During the Twenty-first Century we will sharply reduce our risk for disease, greatly improve our therapies, recalibrate our longevity, modify the plant and animal species that we most depend upon, and begin to enhance our capacity for intelligent action. Human cultures, which (like genes) transmit accumulated information forward in time, are already an established factor in the mystery of evolution. Sometime in the near future (for what is a century in the life of a species?) humans will use biotechnology to orchestrate changes in the gene pool.

Portents of this fascinating, disturbing future abound. We have begun to redesign the crops that sustain us. Each year millions of acres of corn and soybeans are grown with substantially less use of herbicides because a single gene to confer resistance to pests has been added to their genomes. We are inserting genes into farm animals to create an almost unimaginable new kind of factory—one in which rare proteins of great economic value can be extracted from the milk of goats! Thirty years ago, the idea that we might be able to decipher the entire DNA sequence of a bacterium was a fantasy. Today, in the leading genome centers, such a task can be completed in less than a week. Advances in genomic science during the 1990s have transformed the pharmaceutical industry. In 10 years one of its core problems has changed from finding new candidates for drug development to sifting among thousands of potential candidates. In coming decades we will be able to alter life forms in ways that may result in the creation of new species.

New knowledge, when harnessed in the service of capitalism, disseminates at an astounding pace. Most of the tools that enable biotechnology (for example, the polymerase chain reaction) have been developed in the last 20 years. Today, they are routinely used in thousands of companies around the world, as well as in countless college and high school science classrooms. The Western world is, without question, raising a vast army of biotechnologists. There is no sign that the pace of expansion will slacken.

If my musings about the transforming power of biotechnology are even tenuously in touch with reality, then it is important to ponder how its manifold applications will be guided. Who will act as the gatekeepers? Who will lead the way? Who will decide which of the new products that this vast scientific army could develop is most likely to benefit the world? In certain areas, such as the use of recombinant DNA technology to develop new drugs, the regulatory process is relatively sophisticated. In the Western world the U.S. Food and Drug Administration and its counterparts have a respectable history of reasonable oversight. In other areas, such as the rapid, widespread introduction of genetically engineered crops, governmental oversight, even in the most highly developed nations, may not yet be competent either to exercise prior restraint or to engage in the level of monitoring that may be needed to protect people and the environment.

A defining characteristic of our age seems to be that technological advances move with a velocity that defies contemplative, value-oriented analysis concerning the ethics of its implementation. Bioethics, a field that can legitimately claim a history of not much more than 35 years, arose in response to the realization (at first among a handful of theologians such as Paul Ramsey and Joseph Fletcher, physicians like Henry Beecher and Will Gaylin, and philosophers like Daniel Callahan) that we needed to improve our analysis of technological prowess in health care. Driven forward by the first heart transplant in 1967, the expanded use of assisted ventilation in the 1970s, the first successful use of in vitro fertilization to circumvent human infertility in 1978, and increasing use of prenatal diagnosis and genetic testing since the early 1970s, bioethics arose as biomedical ethics. Nurtured in the universities and large medical centers, it matured in the comparative quiet of the academy. But as case after dramatic case (Louise Brown, Karen Ann Quinlan, Barney Clark) commanded the front pages of the press and captured the public attention, biomedical ethicists found themselves chairing blue ribbon government committees, setting policy on the uses of recombinant DNA, writing columns, and learning to be television pundits.

By the 1990s, the decade in which genetics took the center ring, biomedical ethics had, appropriately, morphed into bioethics. The current term reflects our understanding that we have the capacity to influence life in all its wondrous forms and complex communities. Today, many of us realize that profoundly important questions may emerge when biotechnology is harnessed in an attempt to alter conditions far from the hospital.

It has been half a century since the influential British physicist turned novelist, C. P. Snow, worried that the Western Democracies were hampered by the inability of the dominant two world views—the scientific and the humanistic—to sustain a constructive dialog. To some extent, the rise of bioethics is a response to that problem. But throughout most of its young life, bioethics has not engaged in sustained dialog with the captains of the biotechnology industry, the people who imagine the goals, wield the tools, develop the products, and drive them into the marketplace. So long as it was largely an academic enterprise, bioethicists were little different from news analysts. They could make important observations but they constituted an estate far removed from the halls of power.

Only recently, with the growing realization that transgenic crops were rapidly altering world agriculture, that somatic cell nuclear transfer (cloning) was changing the very nature of mammalian reproduction, and that stem cell manipulations were forcing us to reconsider questions about the origin of human life, has it become clear that the biotechnology industry must participate deeply in the discussion of values. Today, unlike just five years ago, bioethicists are advising the biotechnology industry. They deliver seminars, act as consultants, sit on advisory boards, and (rarely) sit on corporate boards. These developments have created a firestorm of debate within the academic bioethics community about the proper ground rules for such activities. This is expected and welcome. Industry needs bioethics, and it needs it now!

This is why *Guiding Icarus: Merging Bioethics with Corporate Interests* by Rahul K. Dhanda is so welcome. If adequate external (read governmental) oversight of biotechnology is not yet in place (and for that matter, even if it is), then it is exceedingly important to enlarge the thinking of those who drive the inventions. Recognizing the shibboleth that the corporate world is motivated solely by profit has never been less true, knowing that corporate executives are for the most part deeply committed to leading principled lives and directing principled businesses, realizing that the captains of the biotechnology industry are acutely aware of the power of the tools at their disposal, and believing that they are up to the challenge of becoming full participants in a sustained public discourse about the ethical uses of biotechnology, Mr. Dhanda has written a book to assist the business world discharge its ethical obligations to the larger society.

His timing is impeccable. In 2002 there are nearly 2000 biotech companies operating in the United States, and at least one new company commences operations every day. A substantial number of these companies engage or will engage in activities (research with human subjects, DNA banking, stem cell research, drug development, and seed modification, to mention a few) that inevitably raise ethical and policy questions. Yet, so far, with a few notable exceptions, the companies have not adequately confronted the questions, nor realized that it is good business both to do so and to do so publicly. Consider, for example, that at a recent national meeting on ethical issues in biomedical research of more than 200 attendees, not one person represented a biotech company, even though those companies collectively perform over half the research!

The Biotechnology Industrial Organization (BIO), the industry's trade organization housed in Washington, D.C., has done a commendable job in speaking out on public policy issues. But, in only a handful of the individual companies, even the large ones, are there internal programs through which the employees and management can explore ethical issues that are relevant to the company's business future. When I talk to biotech executives, I frequently hear them express frustration at the lack of educational materials that they can turn to. This deftly written volume helps fill the void.

No single book could exhaustively examine the entire landscape of business and bioethics, and Dhanda has not undertaken such a quixotic task. Instead, he has chosen four paradigmatic areas—genetically modified crops, DNA data banking, personalized medicine, and stem cell research—as case studies for biotech managers and those (and they are legion) who are observing the industry as it matures. In terms of scale, the changes already evident from the introduction of biotechnology into world agriculture are without equal, as is the magnitude of public unease, especially in Europe. The agricultural biotechnology sector in corporate America has also been subjected to scathing criticism, much of which has been directed at Monsanto. As is often the case in business, careful analysis of its failure—in this case, a failure to engage the American public in discourse about its decision to transfer a gene into a large fraction of "our" corn and soybean crops—offers important lessons to guide corporate actions in the coming years. Dhanda's reflections on the GMO fiasco are reason enough for biotech managers to read this book.

DNA data banking—the storage of tissue or DNA samples, usually with matching clinical and demographic data—is growing at an astounding pace. Over the last three years, venture capitalists have funded a handful of companies that are based on a business plan in which DNA data banks constitute the core asset. Major pharmaceutical companies, such as Pharmacia, believing that such repositories will make drug development more efficacious, have created DNA data banks into which the samples flow as an adjunct to various clinical trials. Realizing the ethical problems that can arise in operating such repositories, Pharmacia also created an external ethics advisory board to monitor and guide it. One of world's largest drug companies, Aventis, is creating a new entity that, if successful, may become the world's largest DNA data bank. From the start, its management team is thinking carefully, as it should, about the ethical issues that suffuse such an effort. Mr. Dhanda competently explores the privacy debate and related issues that surround this activity. In so doing, he explains why industry must be so cautious in these activities.

That in the future healthcare will be keyed to the paradigm of personalized medicine—the use of detailed genetic information about an individual to guide risk assessment, disease prevention, and treatment, including choice of therapy—is indisputable (although the timing of this successful transition is certainly debatable). Of course, personalized medicine can only become reality if people are unafraid that genetic information about health risks will be used against their interests. The fear of genetic discrimination, a concern that dogged developments in clinical genetics throughout the 1990s, seems to have crested, but it has not yet receded. The concern lingers despite the enactment of a federal and many state laws that explicitly protect persons from misuse of genetic information by group health insurers. I doubt that the enactment of still more laws is the answer. What we really need are: 1) new applications of genetic information that the public will recognize as constituting a powerful advance in health care, and 2) a discourse in which members of the biotech business world inform the public how exactly such knowledge will be used. Dhanda's insights into the future of personalized medicine are a must for those who are working at companies that have hitched their wagons to that star.

Although relatively few biotech companies are dedicated to stem cell research, Dhanda wisely chose this as one of his case studies. Although the public debate about stem cells sometimes seems confused and inarticulate, it bears close review because it reveals a core uneasiness that lay persons feel about the entire biotech industry. The stem cell debate is on the surface a controversy that pits fundamentalists from the pro-life (antiabortion) camp against the secular relativists who dominate technological societies. But the controversy would not have become so heated if there were not more. Many of those who would identify themselves as supporters of biotechnology perceive that stem cell research, which is closely related to cloning technology, approaches an aspect of our humanity—how we reproduce—that is, if not sacred, sacrosanct. They sense that in its broad applications, stem cell technology will dehumanize us, commodify us, knock us down yet another rung on the ladder of creation from which we have been steadily descending for two centuries. Dhanda helps by clearly explicating the debate and suggesting how to engage the public about it.

During its 35 year history, bioethics in the United States has made many important contributions concerning the thoughtful and measured use of astounding new technologies. But only recently has it begun to move out of a largely academic existence to engage and be engaged by the world of business. There is a palpable need for business to explore the issues Dhanda raises. *Guiding Icarus* offers a flight plan; it will help business leaders in biotechnology to chart a course from their treasure laden islands to the continents they wish to serve. This book will do much to foster a greatly needed discourse and by so doing will raise both public understanding and esteem for biotechnology.

PHILIP R. REILLY

Concord, Massachusetts

PREFACE

. . . so then to unimagined arts
He set his mind and altered nature's laws
—Ovid, *Metamorphoses*

Biotechnology is mythic. Granted, it is the tool of man, but it accomplishes what myths and legends could only realize in the realm of imagination. For instance, chimeras are no longer the products of storytellers; they are the creation of scientists. By melding the genes of different organisms, biotechnologists are coming closer to these hybrid animals conjured by poets, balladeers, and troubadours of old. Centaurs seem unlikely in the near term, but the minotaur may be closer, as researchers at one company have already placed a human nucleus into a bovine embryo. Biotechnology's accomplishments seem as likely a result of the feats and follies of gods and heroes as they are of scientists. And whereas we recognize the reality of the latter, lessons from mythology should not be ignored. If we forget our myths and dwell in the rational world of science, we may sacrifice the awe inspired by these tales, running the risk of ignoring the prescient warnings that pervade them. After all, explaining the inexplicable was only half the function of myths.

There are as many ways to examine biotechnology as there are disciplines that claim intellectual sovereignty over it. While this book centers on the oft-discussed ethics, or more precisely the bioethics of life science research, its place in this literature is unique. Of the multitude of examinations, few, if any, have focused on fitting bioethics within a corporate context. Because of this deficiency, no one has satisfactorily reached out to those people on both sides of the corporate fence who need to engage these debates: corporate representatives and the public that is affected by these technologies. Another result, then, is that few have offered constructive guidance to industry as to how it can address bioethical issues; even fewer have explained how a bioethics-aware corporate strategy can aid companies to meet their economic and social goals. While searching for a book that did this, I found that it had yet to be written.

Before continuing my discussion of the text's background, I must disclose something. In Chapter 4, Personalized Medicine, I use Interleukin Genetics as my case study. At the time of this writing, I am an employee of Interleukin Genetics, and I feel it necessary to bring this to the reader's attention. I leave it to the reader to decide on the merits of the company as an ethical agent, but I caution him or her against too much skepticism when it comes to the lessons offered in the chapter. It was frankly easier for me to study a company I know well. The "Recommendations" section is composed without specific reference to any companies, so any doubt as to the veracity of my analysis should not obscure the importance of that section. I should also note that one of the reasons I chose Interleukin Genetics as an employer had to do with their commitment to exploring the ethical dimensions of genetic research.

Philosophers have asked many questions about technology, which have been echoed, improved upon, and effectively explored by bioethicists, but, with few exceptions, these dialogues have remained exclusively among and accessible to these specialists. Others can engage them, as writing this book has required of me, but the dedication necessary to build the tools to sufficiently do so demands more effort than, say, a CEO has time to apply. One must learn the philosophical foundation, the history of science and medicine, and the applied tools that are currently specific to the academic realm. Scientists have also looked at many bioethical issues, and often their response has been to clarify and intensify their research; consequently, healthcare improvements will derive substantial benefits from these scientific investigations. When these experts weigh in on the public concerns, the focus is often on explaining the science, as the logic usually follows the pattern: "If the public only knew what I know, then they would recognize the imperative to do this research." Even so, the discussion can be too technical to reach the public or industry representatives. Furthermore, scientists are personally invested in their field, and their dedication to it casts some suspicion on their motivations. After all, just because the benefits are clear to the technology's purveyor does not mean that it translates to the public. Finally, the business component of biotechnology almost exclusively focuses on how to maximize profit and leverage technologies to that end. Analyses in this category reflect not only this focus, but also the distance many of these ideas have from science and moral philosophy. That is not to say that moral philosophy has not found a place in the corporate context. Business ethics are important (represented by an expansion of the literature and increase in subscriptions to related courses), but it is not bioethics. Although all of these areas—science, philosophy, and business—overlap in the realm of biotechnology, they are often addressed separately. In more progressive cases, two topics may meet (e.g., the aforementioned business ethics, or history/philosophy of science, or science studies), but never have all three been systematically explored together. After all, they are different fields, with different vernaculars, but that should not give cause to ignore their symbiosis. In response, I have aimed to act as a translator examining this congruence.

Following my invocation of the mythic nature of biotechnology, one may respond by asking which tales should we heed? Let us concentrate on the ancient Greek and Roman myths, although their anecdotal counterparts exist in any number of cultures. One that frequently comes to mind when discussing the untold promise and peril of biotechnology is Pandora's Box. Indeed, many texts discussing biotechnology, and even science and technology in general, have relied on this powerful allusion to explain their arguments. Surely, we should not open the box that seems to hold as much mystery as the promise of harm. Yet, we already have. We alter genes to try to cure disease and create martial germs; we tinker with nuclear power and weapons; we capture light in wires to trade in information or obtain it from others. We have opened that box, but we have learned to harness nature to help more often than to harm. Pandora's story has a lot to offer biotechnology, as it has in any number of expositions on the topic. Although elements of it penetrate this book, it is not the story that I will tell.

Let us not forget Prometheus. Mary Shelly used this image to great effect in her novel *Frankenstein.* Its subtitle, *Prometheus Unbound,* reveals the peril of "improving" nature. One may believe one can create life or alter it, but blindly applying the knowledge may have severe consequences—it may not be a life that is created; it may be a monster. As those familiar with the myth of this titan know, Prometheus created the form of man with clay, which later had life breathed into it by Athena. After creating this organism, Prometheus delivered a powerful technology to humans, that of fire. Yet, by placing this liberating flame in the hands of the Gods' subjects, Prometheus invited the divine fury of Zeus. Chained to a mountain upon Zeus' command, Prometheus would have his flesh torn daily by a vulture that would visit him as soon as his wounds had healed. Spreading technology seemed like such a good idea at the time. This story brings us closer to the context of this book. Like Prometheus, humanity now has the power through biological science to recreate its image, molding DNA rather than clay, but one must be cognizant of the Promethean potential of such actions. At the same time, a new technology has been placed in the hands of practitioners that may lead to repercussions beyond the anticipation of those working with it. Even more relevant to industry, a group may benefit from the power given to it by an outsider (corporations), but the deliverer may suffer. And unlike Prometheus' situation, more groups than fickle Gods can judge and punish—society, industry, and government represent just a general few. This particular tale is dense with meaning, and this short paragraph only begins to scratch the surface of its counsel. Although it is relevant to the argument within these pages, Prometheus' biography still does not completely carry the reader into the full context of this book.

As is no surprise given the title of the text, it might seem that Icarus assumes the exemplary role of the essay. But he does not do this alone. Icarus, if you recall, was the "hero" whose hubris brought him tumbling to his death. With artificial wings fastened to his back by wax, Icarus took flight to free himself from the isle of Crete, where the ocean formed the bars of his prison. With this new technology—the wings built by his father Daedalus—Icarus became carried away, literally and figuratively. He never doubted his control and understanding of this technology; guided by pride, he flew too close to the sun. With the solar heat intensifying as he rose, his wings became unfastened, and he fell to his death into the sea that has since borne his name. Lamenting his loss, as a father would, Daedalus believed that he did all he could to free them and stave off this outcome. Indeed, the description that opens this preface is taken from Ovid's account of the tragedy, and they describe what Daedalus had to do to save himself and his son. Not unlike some descriptions of biotechnology, Daedalus set his mind to "unimagined arts . . . and altered nature's laws." But he did more than that. Seasoned by his mistakes, Daedalus knew that new technologies bring unimagined perils, and thus he warned his son:

> "Take care," he said
> "To fly a middle course, lest if you sink
> Too low the waves may weight your feathers; if
> Too high, the heat may burn them. Fly half-way
> Between the two. And do not watch the stars,
> The Great Bear or the Wagoner or Orion,
> With his drawn sword, to steer by. Set your course
> Where I shall lead." . . .

Daedalus led, and Icarus strayed. Daedalus lived, and Icarus perished. Who of these two should be the model for life science?

In my view, most life science companies are in very much the same position as Icarus. They work with powerful new tools, but with limited guidance as to how they can hurt or help stakeholders. The belief is: Develop the science into a product, and prosperity will follow. Granted, realization of these goals is significantly more sophisticated, as it involves clinical trials, experimentation, manufacturing, marketing, distribution, and many other areas of expertise. Yet, if guided solely by their belief in this teleological pattern, these companies run the risk of perfecting their technology at the expense of their own, and their customers' interests. These new technologies have predictable and unpredictable outcomes. Luckily, bioethics tries to anticipate as many of the implications of biotechnology as possible. The next step is to apply these ideas to the corporate setting.

The story of Icarus and Daedalus offers great insight to the life science industry. Icarus made a choice, and part of that choice was to ignore the advice of Daedalus. Companies have that same choice; they may *hear* words of advice, but the choice is theirs to *listen* or not. Possessing technologies that may imperil themselves and others, firms must appreciate the gravity of the associated responsibility. However, warnings need to be more effective than those offered by Daedalus. Although he kept his own counsel, recognizing the twin dangers of fire and water, he did not impress this danger upon Icarus. While you read this book, I hope that my words make a more convincing case. And as companies learn more about their technology's place in society, they may gain the wisdom of Daedalus. The choice is theirs: lead, follow, or stray; survive or perish. With this in mind, I hope that I do a better job than Daedalus at guiding Icarus.

RAHUL K. DHANDA

Cambridge, Massachusetts
February 2002

ACKNOWLEDGMENTS

Although this is my first book, the experience has left me with the deep sense that texts cannot be written without a tremendous amount of both intellectual and emotional support. I am very lucky to have had ample amounts of both supplied by family, friends and colleagues.

I cannot put into words how grateful I am for the encouragement offered by Phil Reilly, whose contribution to this text extends beyond its foreword to its critical review and many thoughtful discussions regarding its composition. I would also like to thank Paul Rabinow for guidance that came well before I decided to write this book, but influenced much of the focus and direction of the text.

Many others offered both personal and financial support, a good deal of whom are family. This same group of people inspired me to write this book. More precisely, my fear that abuses of biotechnology may someday damage them drove this project, and as Iris Murdoch once observed, to truly understand an intellectual, one must understand his fear. First, I would like to thank my brother Samir "Bob" Dhanda, who provided me with a great deal of technical support. I would also like to acknowledge the support of my dear friends Jeremy Pienik and Rachel Kaplan, whose training in sociology proved invaluable. I extend a special thanks to Jeremy for over two decades of dedicated friendship predicated upon our mutual intellectual development. I am indebted to Greg and Michele Sabatino, the former for his thoughtful discussions regarding issues surrounding privacy and the Internet, and the latter for the concise thinking that led her to conceive the book's final title. For his intellectual curiosity and queries into the substance of the text, I am grateful to Greg Lalas, without whom I would not have settled on Icarus (or more precisely Daedalus) as the book's prevailing allusion. My colleague, who splits his time amongst the roles of "boss," friend, and mentor, Paul "Kip" Martha, critically reviewed portions of the text, and substantially improved Chapter 4, for which I am very grateful. My close friend and colleague, Chuck Scott, deserves recognition for his review of the entire manuscript. I would also like to thank K. C. Cargill for poring over many ideas with me during our frequent interactions. My brother Anish Dhanda and his wife Stefanie Dhanda provided more than just encouragement. Their belief in the importance of this text and my ability to write it translated invaluably into both financial and intellectual support; yet I am most grateful for the sterling professional examples they set. Anish, in particular, has always impressed upon me how one's conduct, professional and personal, should always mirror one's ethics. My mother, Shashi Dhanda, deserves special thanks; she very patiently hosted me for some months while I worked on portions of this text. There are no words that a son can use to adequately acknowledge the contributions to his life that a caring mother makes, so I hope she understands that my uncharacteristic lack of words indicates an abundance rather than shortage of gratitude.

Others deserving of recognition are Kiril Stefan Alexandrov, Mark Audeh, Paul Billings, Christianne "Chrissy" Brissette, the crew at Chez Henri, Peter Courossi, Frank DeFazio, Doug and Teresa Delaney, my colleagues at Interleukin Genetics, Noelle Gracy, Patricia Nelson, Jose "Pepe" Pineiro, Jody Ranck, Fred Simmons, Lee and Marci Lerner, Paul O'Connell, Jim Stanek, Lisa Van Horn, Neal Wadhera, Priya Wadhera, and Ike Williams.

I would like to extend a special thanks to Luna Han, my editor at John Wiley & Sons, who recognized the great need to expand the bioethics/ biotechnology literature to include analyses of industry's role in this complex equation. In general, I would like to thank Wiley for increasing, through their publications, awareness of the complex, beneficent, and potentially dangerous relationship between science, society, and industry.

There are two people who deserve special thanks because they are substantially responsible for the book taking the form that it has. First, I would like to thank Rick Onorato for contributing the illustrations. With very few exceptions, his talented hands created every picture in this text, making this book much better than it would have been with words alone. Finally, every word in this text has been reviewed multiple times by Peter Causton. Peter acted as my own personal researcher and discussant from before I put a single word on paper through the submission of the final manuscript. I can only hope to return the favor when he begins working on his own book, which I eagerly anticipate.

Although all of these individuals contributed to the strength of this text, I am solely responsible for any errors that may exist within it.

R. K. D.

1
INTRODUCTION

> Increasingly, knowledge about the genome is becoming an element in the relation between individuals and institutions, generally adding to the power of institutions over individuals.
> —Richard Lewontin, *Biology as Ideology*

Biotechnology has and will continue to change the world by altering our most basic conceptions of what it means to be human. Deservingly, this tendency has led to a substantial critique warning against the danger of such a pervasive set of technologies. Countless volumes ranging from the most rigorous academic works to popular science open with some variation of the phrase "The great promise of technology X brings with it many social questions. . . ." Unfortunately, the truth behind this statement is sometimes lost in the cynicism that has made it cliché. Nonetheless, it is true in almost every application. Technology changes society—whether you are a Marxist who sees its aim as a means of liberation, a capitalist who sees it as a means to profit, or a biologist who sees it as part and parcel of the human evolutionary prerogative. Does it change it for the better or for the worse? Despite the position of Luddites or their counterparts, technophiles, technology can and will be used for conflicting ends. In the same way that any institution can be abused, technology can be misused. Religion, for instance, has always claimed to be a measure of moral (good) behavior, but many perform contestable acts in the name of God for what seem to be questionable reasons. Technology and science, however, tend to have a more threatening place in society because, unlike the institution of religion, they are not recognized as pursuing inherently noble objectives. Advances in technology are often considered alongside their potential harm; this may seem to be an unfair criticism of technological progress, but difficult to object to when weighing progress against the human interests that technology is meant to serve.

Although there are many technological realms that deserve attention, this book focuses on a specific one: corporate biotechnology—that is, biotechnology performed in the private sector. The private and the public sectors deserve different types of attention because they mediate technology and society in different ways. Public research, when it truly is devoid of commercial intent, proceeds to deliver therapeutics to the public as a means of public service. The private sector also intends to aid public health, but through different mechanisms. Furthermore, corporations deliver more health-

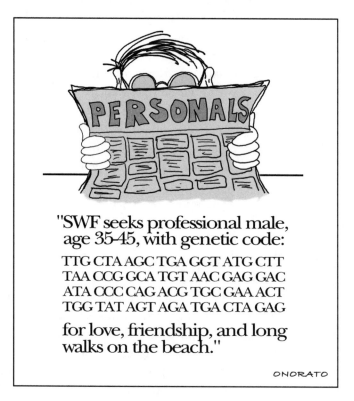

Figure 1-1.

related technologies than academic centers. Even when academic researchers create some type of breakthrough, a private corporation usually assumes the licensing rights and manufactures the new drug or diagnostic. Medical technologies are delivered through the private sector; thus, any discussion of bioethics must account for the unique characteristics of these institutions.

In another book that addresses this topic, *French DNA*, anthropologist Paul Rabinow explains that there are new assemblages forming around biotechnology that deserve great attention:

> Such a common place, a practiced site, eruptive and changing yet strangely slack, is filled with talk of good and evil, illness and health, spirit and flesh. . . . Amid all the discord, however, all parties agree that the future is at stake, and that there is a pressing obligation to do something about it.[1]

Biotechnology and life science touch upon our most basic understanding of ourselves; with every new discovery, we are removing the uncertainties that surround our place in the world, and we replace these notions with a less mystified view of our very nature. Given the focus of this book, one might wonder "why 'corporate' biotechnology and not all of biotechnology?" My answer is "because there are few, if any, books that address corporations specifically, and none that offer corporations a practical means of confronting the ethics of their research." That is not to say that academics have nothing to gain from this text; on the contrary, because these pages explore how corporations can integrate bioethics into their operations, bioethicists stand to gain by learning how best to apply their ideas to industry. At present there is a disparity between the form of advice offered by most bioethicists and the adaptability of that advice to industry. Although not geared towards academic bioethicists, perhaps this text will help the discipline recognize how it might more effectively address the specialized conflicts between bioethics and the private sector. Until this shift occurs, this book hopes to end this disparity by appealing to industry in a way that bioethicists often fail to do.

---

[1] Rabinow, Paul. *French DNA.* Chicago: University of Chicago Press, 1999, pp. 4–5.

Corporations are unlike other institutions. They do not speak the language of academic bioethicists and, for that reason, they may hear what academics have to say, but they may not understand how that critique fits within their day-to-day operations. The converse is true, as well; although many corporate representatives of biotechnology firms have as much interest in bioethics as bioethicists do, a corporate officer will have a difficult time inviting bioethicists to engage in the rhetoric of business. Granted, ethics are something that each person, regardless of training, can appreciate; however, the subset of knowledge that constitutes bioethics is removed from everyday ethics; it consists of foundational principles like autonomy (respect of an individual's choice over his or her medical decisions), beneficence (the obligation of scientific practitioners to work in the best interests of patients or their customers), nonmalfeasance (the obligation not to harm individuals), and justice (the obligation to work toward society's interests). These ideas are useful tools for any individual, but they do not frame most people's everyday conduct.

It would be incorrect to interpret this book as an effort to belittle the work of bioethicists; indeed, it is their hard work and insights that are being directly appropriated in this text. Philosophers, sociologists, and other scholars of the humanities and social sciences have influenced this essay as well, and for that reason it must be said that a good deal of the information in this book is not novel. Rather, the novelty lies in combining many different disciplines to make these ideas accessible to the scientists and corporate officers who understand the challenge of altering the bureaucracies that frame private industry. Although biotech firms are known for their freewheeling management styles, the more successful they become, the more bureaucratic they will have to be. It may seem restraining, but it is a necessary shift for maintaining operations on a large scale. To see how bioethics fits within the corporate context, then, the application of these ideas needs to be examined within it. Medical anthropologist Arthur Kleinman has pointedly observed that cultural differences influence the value systems surrounding biological science[2]; scientists, corporations and ethicists represent their own cultural

[2]Kleinman, Arthur. Moral Experience and Ethical Reflection: Can Ethnography Reconcile Them? A Quandry for the New Bioethics. *Daedulus, 128*(4): 69–97, 1999.

form, and their values correspond to this form. That is not to say these different cultural systems are exclusive of each other or that they cannot inform each other. Quite the contrary, if ethics are to be effectively introduced into corporate biotechnology, a bridge must be drawn across the different cultures that affect scientific practice, and the individuals who build such an edifice need to be informed by each culture, highlighting a plurality of voices. The goal of this text is to erect enough of that bridge to convince biotechnology that it is worthwhile, safe, and necessary to cross it.

While each chapter discusses specific technologies and highlights their benefits, the major emphasis is placed on introducing the ethical issues, while offering recommendations on how to approach them. The primary message conveyed in this text is the following: it is not enough for a company to have faith in its technology anymore; the company must also be responsible for it, and responsible to the many stakeholders who are affected by the progress of biotechnology. Furthermore, the definition of stakeholders has to be broadened. Biotechnology is too pervasive for industry to err on the side of exclusion when identifying stakeholders. Although it may seem as simple as realizing that we are all affected by these technologies, industry insider or not, it is not as simple as believing that we are all in this together. Too often, industry begins its research from the perspective that cures outweigh potential harms. Activists and other critics focus on the risk of harm, and elevate the concept of risk over that of therapy. These are different world views, and the biotech sector needs to learn that other perspectives may not be strong enough to stop the progress of science, but strong enough to cripple its widespread acceptance. What CEO of any company, acting as just a human being for a moment, would choose to donate his or her DNA to a research effort at the expense of his or her family's medical insurance? What head of research would feed genetically modified food to his or her children if the data to support its utter and complete safety was questionable? For those who would do both, what convinces them that they have the right to put their family at risk based on belief and knowledge that is not easily shared with those affected? However, if these questions have not been asked and explored, if the safety of these actors and their loved ones have not been examined, why wouldn't an executive unleash the technology on the rest of the world? For these reasons, creating the most promising technology, whether supported by the most vocal opponents or not, does not excuse doing so at the cost of social responsibility. As biotechnology progresses, it must do so hand in hand with ethics.

It is here that disclosure on my part ought to be given. This book is not meant to be a bioethicist's analysis. The experience and attitudes that are presented in the text are informed by bioethics, but they are written from the point of view of an industry insider. Of the voices chosen, the dominant one is meant to be that of a biotechnology sympathizer, yet one who has filtered the bioethics literature to make it accessible to individuals working in this field. To build on the previous metaphor, the bridge that this book is building may be closing the chasm between bioethics and private biotechnology, but it is beginning from the side of the former. However, it is a joint effort; thus, the tools and skills are derived from the latter. In the end, each side will do more than just see the other; they can reach the other by crossing the bridge, and come back just as easily, but, hopefully, more enlightened.

This disclosure explaining my sympathies is meant to do more than just explain the filter by which this book is written and with which it should be read. I offer it as a cautionary note for industry and a preemptive measure against some critics. It should be read as a cautionary note because I clearly favor the technologies discussed in the text, and thus the reader should avoid the temptation of believing too strongly (or blindly) in the unchecked promise of technology; it can also be a preemptive note because that sympathy is not to be mistaken as apologia or justification for pursuing research at all costs. It should not matter whether the discussions regarding the technologies favor industry. To effectively engage the debate, it does a company no good to turn a blind eye to critics and pay special attention only to those who show great faith in scientific progress. Like the animal stalked by the tiger, the problem cannot be eliminated by closing one's eyes. Choosing ignorance is not a defense—indeed it may be fatal. Furthermore, bioethical advice to a company is no different than any proffered advice: telling the corporation what it wants to hear rather than what it needs to hear will only end up damaging the firm. What companies want from bioethicists is the information that the firms *need* to hear. To tell them that their research is without ethical complications when it truly is, results in a disservice to both the discipline of bioethics and the company. Indeed, in a recent article in *The Hastings Center Report,* bioethicist Carl Elliot explains his reservations regarding the melding of bioethics and industry:

> My concern is that bioethics as an enterprise is separating itself from a particular conception of what the moral life is about. . . . The danger of living in this vast corporate wilderness is that someday there will be no rebellion, no protest, no dissent that has not been bought and sold. I worry that each corporate check cashed takes us one step closer to the notion of ethics as a commodity, a series of canned lectures, white papers, and consultation services to be purchased by the highest bidder and itemized on an annual budget report.[3]

In short, Elliot is concerned that corporate involvement might rob bioethicists of their disciplinary integrity. Coincidentally, any company that would invite bioethical advice shares the same worry; it is exactly staying true to the objective critique of bioethics that gives this knowledge the necessary weight to make it useful to industry. If bioethicists believe that they are hired to play apologists, then both industry and the discipline are wronged. Corporations know where they stand, and they do not need to pay someone, regardless of their credentials, to be reminded of their position; instead, they need to be informed whether their footing is solid or not. What is a more legitimate concern is misappropriating the critique and using it to identify which areas to hide rather than improve. In a similar vein, bioethics is not a public relations tool, which is not to say that it ought not be used in PR. In some cases, as will be seen later in the text, public relations is the best sphere to bring issues to higher social awareness. Furthermore, companies that are adopting such measures can use their marketing divisions to reform their entire industry by showing by example how to conduct a responsible business.

If used correctly, bioethics can become a seamless part of the corporate landscape. Standard operating procedures and corporate culture will each re-

---

[3]Elliot, Carl. Throwing a Bone to the Watchdog. *Hastings Center Report, 31*(2): 10, 2001.

quire reconception, and understanding how these factors work together with bioethical ideas is the most difficult challenge facing this transformation. This is the problem that this text tackles; however, recognizing this need and wanting to achieve this goal are not one and the same. Both steps need to be taken, and the next obstacle facing this work is convincing those people who believe that corporate viability and ethical dedication are mutually exclusive. Many companies proceed bioethically because they believe in doing the right thing, and they have prospered by it. As far as defining their operations, their core values are consistent with being socially responsible. Genzyme, for instance, has an officer dedicated to recognizing bioethical issues. Millennium Pharmaceuticals' former Chief Business Officer Steve Holtzman played a similar role at his company, but he and Millennium have even gone a step further. Holtzman is the only industry representative to have joined the National Bioethics Advisory Commission, and, furthermore, he has forgone involving himself on the boards of other biotech companies because he "would rather spend [his] non-Millennium time working on industry and its attitudes towards bioethics."

Because these examples are not always convincing, more specific examples of corporate benefits will be presented in the text, and, perhaps more relevantly, ways of avoiding damage by acknowledging rather than ignoring bioethics will be discussed. While this may be the only book directed at guiding an industry toward adopting bioethics, it is not alone in trying to convince the industry to make this shift. Aside from academic attempts, industry experts are already pointing out the necessity of corporate approaches to bioethics. As stated in a recent report by PriceWaterhouseCoopers entitled *HealthCast 2010:* "The Human Genome Project will push the envelope in terms of how medical information is collected, disseminated and organized. In addition, new waves of medical devices and drugs will elevate questions of medical necessity, personal responsibility and rationing." The report offers the following action items:

- Establish a bioethical framework in which to make decisions
- Consult with established bioethics departments at other institutions[4]

---

[4]PriceWaterhouseCoopers. Implication #12. *Healthcast 2010. http://www.pwchealth.com/ healthcast2010/.*

These efforts are not meant to restrict research or limit the progress of biotechnology. Indeed, they are intended to facilitate research because they address the always-changing social conceptions of biotechnology. When the public has reservations over research, then at least one stakeholder (representing a large number of consumers in most cases) will be alienated if the issues are not addressed. In fact, the following repercussions may ensue—indeed many already have, as will be evidenced in the later chapters—if biotechnology does not incorporate sensitivity to ethical questions into its daily functioning:

- Loss of consumer interest and loyalty
- Litigation target (justified or not)
- Damaged public image
- Competitive disadvantage
- Hamper collaborations
- Threaten mergers and acquisitions
- Fall behind legislative and regulatory trends
- Increase employee turnover
- Decrease market capitalization
- Associations with a type of industry and its negative connotations

Although all of these issues highlight the potential damage associated with avoiding bioethics, adopting ethical strategies can offer a competitive advantage instead of injury in the following ways:

- Expand consumer/customer base
- Increase consumer loyalty through branding
- Lower insurance premiums
- Proactive/preventative reputation with regulatory agencies
- Recognize new market opportunities
- Avoid adverse events/recalls
- Attract collaborators
- Aid in passing diligence for Mergers and Acquisitions
- Anticipate legislative and regulatory trends
- Improve public image
- Increase employee pride

- Reduce turnover
- Reduce legal costs

These are all compelling reasons to adopt an ethical framework for research, but they still do not replace the best reason of all, which is that the company *wants* to be bioethical.

There remain two more general questions about the text that require clarification before proceeding to the next chapter. They center on who and how. For whom is this book written and how is it to be used? To answer the first question, this book is aimed at a diverse group of people within a narrow range. The range is, as stated earlier, anyone interested in biotechnology, although it is primarily structured to aid industry representatives. The diversity lies in the many people who can use this book. Because the text deals with the confluence of bioethics, biotechnology, and business, it is written so those from each sector, ranging from corporate executives to graduate students in the biosciences to bioethicists to business school students, can engage the entire text. However, its form is such that it will not only explain how to adopt corporate approaches to bioethics, but also aid industry's decision makers in implementing them. Although it is intended for the desks of corporate officers, it is also targeted at regulatory affairs managers and vice presidents. By reading this book, corporate officers will better understand the issues facing their companies, whereas regulatory affairs managers, as well as individuals working in quality systems, are best positioned to use the tools presented here to improve their standard operating procedures. Because many of the recommendations are geared toward specific departments (e.g., marketing, research and development, etc.), the vice presidents responsible for those particular divisions will also benefit from this information. Each chapter is divided into different sections, and one need only read those aspects of interest, although the best vantage will be gained from reading the book in its entirety.

The book's structure describes how it can best be used. Each chapter is divided into six sections: a brief introduction, an explanation of the science, an exploration of the benefits, a description of the ethical issues surrounding the technology in question, a discussion of case studies focusing on how industry groups have responded, and recommendations to address the ethical concerns. Depending on the reader's familiarity with the subject matter, the explanation about the science and its benefits can be passed over. These sections are merely meant to set the discussion for the ethical issues, but they are very important, as the technological points and their benefits are almost always at the foundation of debate. For readers who do not work in the field, these sections are critical to understanding the remainder of each chapter. The ethical issues are the heart of the book, thus they deserve special attention. Most of these arguments are derived from a detailed survey of the literature and the industry to see which issues have risen to the top of public debate. These are usually the points that most stakeholders, be they activists or investors, have recognized as having major implications for society, and these are the first concerns that require confrontation. It would do the field a disservice to attempt to address every issue facing each technology in this limited space, but a suggested reading list is offered in the Appendix for those with unanswered questions. The penultimate section discusses the success or failure of companies to deal with these issues, while explaining the repercussions for each company due to its foresight or blindness. The final section lays out practical recommendations that biotechnology firms can adopt and integrate in an effort to become more bioethically aware.

The technologies presented in this text are purposefully varied to represent as broad a range as possible. There are too many specific technologies being developed in the life science sector to present them all, but the following were chosen because they offer a model for many similar technologies or ethical issues. More precisely, these technologies may become widely accepted, and the ethical issues may be mitigated at some point in the future; however, new technologies will develop, and they will most likely tread on similar ethical ground, requiring the same lessons learned from the technologies discussed in this book. The specific issues in the text, by chapter, are: Chapter 2—Genetically Modified Food, Chapter 3—DNA data banking, Chapter 4—Personalized Medicine, and Chapter 5—Stem Cell Research. Representing various degrees of controversy, these areas were chosen as much for their bioethical concerns as their level of scientific advancement. They are all in use today, and businesses based on these fields are only increasing in number.

Chapter 2 focuses on the modern, paradigmatic case of biotechnology and bioethics. Genetically modified foods have taken center stage in the global debate over biotechnology. Because the science is so pervasive, affecting a ritual considered sacred all over the world, that of consuming food, this technology was destined for scrutiny. Indicating how pride can fell a major program, Monsanto's Icarian woes are used as the example of what not to do when presented with strong opposition from activists. Although Monsanto has taken significant steps towards being a bioethically aware company, its prior attitude of knowing better than its critics damaged not only itself, but the entire genetically modified food industry. In this case, policies, labeling strategies, information management, and transparency are recommended as a means of integrating bioethics into corporate strategy.

In Chapter 3, DNA data banking is used as a platform to discuss an issue that is extending beyond health care and into information technology: privacy. Because DNA data banks amass substantial amounts of data related to patient health, institutions outside of biotechnology will be interested in the information stored in these warehouses. Employers and insurers will want to access the data, and may possibly discriminate against individuals in the future based on their genes. Using DNA Sciences as a case study, methods of avoiding privacy violations are explored. While a number of corporate approaches are offered to maintain the donor's confidentiality, the chapter centers on expanding the precepts of informed consent, which is already a standard procedure in most research organizations.

Chapter 4 focuses on the developing field of personalized medicine, using Interleukin Genetics as an example. In this section of the text, many of the issues from Chapter 3 are relevant, but the utility and accuracy of screening individuals for disease is also specifically explored. The psychological damage associated with gaining knowledge about a condition that may have no cure is weighed against the research imperative to increase discovery. Among the recommendations in this chapter, the place of genetic counselors in biotechnology is explored. Although this is one application of pharmacogenetics, the field is much broader than the traditional conceptions of genetic testing, which primarily focus on diagnosing disease. Pharmacogenetics reveals that a more powerful use of genetic information is to predict specific therapies and drug treatments. Some drugs will work better in one population, and another may serve the remaining patients. Although many large pharmaceutical companies support this field in principle, targeting *specific* individuals within a disease population rather than the *entire* disease population pits the power of genomics against the drive for profit. Pharmacogenetic companies are placed in a unique position wherein the interests of the patient and the interests of the company are one and the same. These firms want to popularize the assays that hone therapies by predicting drug response, which will usher in the era of personalized medicine; however, this revolution requires that a small, new industry (pharmacogenetics) convince a large, old one (pharmaceuticals) that the ethical shift toward personalized medicine outweighs the profit motive behind pushing one drug when it is not the best therapy.

In the fifth chapter of the text, the highly controversial area of embryonic stem cell research is confronted. These cells offer the tremendous potential to cure many degenerative diseases, as well as other illnesses once thought incurable; however, the cells are derived from embryos in a process that results in their death. Linked closely to the abortion debate, this technology has been gaining more and more prominence in public forums as changing political attitudes continually alter the status of this science in the United States' research agenda. The right-to-life movement and the pro-choice movement use the issue as another sounding board for their ideologies, and the companies working in this field are caught up in a debate that may impede their success. In this case, the ethical issue over abortion is very far from resolution, but the chapter explores areas of compromise regarding the source of tissue used in experimentation that may obviate some of the most pressing problems.

Looking at these technologies, it may seem that they were chosen solely because of their timeliness and the attention they have received and promise to receive over the coming years. However, this was just one reason for their inclusion. The other rationale for including these technologies lies in the types of social concerns with which they are associated. To look at the root of critique, we find that unease stems from concepts of privacy, discrimination, health risks, and abortion/woman's right to choose, rather than the technology itself. The world always asks for cures to be delivered by medical science, and these technologies are in high demand, but wariness is only introduced when these therapies seem to come with catches attached. When looking at these particular social anxieties, those familiar with the bioethical terrain will recognize that they have been problematic in the past with other, now commonplace, technologies. For instance, before privacy was mentioned in the same critical breath as DNA data banking, it was an issue in storing health records, which is a reemergent concern due to the linking of many different databases via the Internet. The abortion question, particularly as it relates to embryonic research, met with heated debate in the 1960s and 1970s as the push for in vitro fertilization was strengthened by the success of the technology. These are just a few examples that illustrate an important point: new technologies will always arise, and even as some of the more polemical become naturalized as a standard health practice, many will be saddled with the same bioethical baggage that seems to be intimately related to the field. Thus, even as the technologies in this text become outdated or accepted, the strategies, advice, and recommendations in this book will still be relevant for new, analogous cases.

Building on this explanation of root causes, there are two final points that deserve attention. Although it contends that firms can adopt tools that make the company more bioethically aware, it is not the intention of this book to outline how the life science industry can solve these problems alone. These concerns center on social questions, and biotechnology is just one example of the many institutions contending with social ills. In this vein, when they focus on biotechnology as responsible for both the cause of and finding the cure for these ethical issues, I will go so far as to say that critics have treated biotechnology somewhat unfairly over the years. Granted, some groups are less bioethical than others, but the pressure to *solve* the issues that pervade these technologies is sometimes unduly applied to this industry. If one were to analyze the history of discrimination in the world, or privacy violations, it would quickly be recognized that these concepts existed millennia before modern biotechnology gave the topics new grounds for discussion. Yet, when bioethics (more precisely, the critiques addressed in this book, and their ilk) is used as a justification to single out this industry as socially caustic, then the critics are doing more damage than good by ignoring some very basic facts. Most importantly, the majority of these issues, such as discrimination, were artifacts of society long before they were symptoms of business. These social evils ought to be the paramount case in discussion, with the technologies or social attitudes that "foster" them considered as examples of their application. To overcome the technology is an empty victory if the root of the problem persists. In many cases, the word "bioethical" needs to be broken down and better analyzed. "Bio" is merely a qualifier for "ethical." It is ethics that deserve the greatest attention, and the "bio" part of it just explains one means of addressing the ethical dimension. To privilege the biological or technical component is to miss the goal that this field has to offer, which is to outline a mode of conduct that is consistent with a moral world view. Just because industry is an easy target does not mean it deserves displaced blame; however, industry must still recognize that it has an obligation to develop and act consistently within a moral framework. It has inherited a set of problems that must be confronted so as not to exacerbate social problems, regardless of who takes the lead in trying to solve them.

Holding industry culpable for problems that lie deep within society may also lead to an even greater, unfair expectation. Some also believe that it is the duty of industry to solve these problems. The debate over abortion, for instance, is so deeply divisive in society that it is absurd to expect the business world to resolve a question that some of the greatest scholars in the world have yet to unravel. This is not an attitude held by most bioethicists, but some of the more extreme critics become so entangled in their antagonism against the industry that they believe that businesses ought to stop their operations because that is the only way to resolve these issues. A better framework for understanding industry's place in these debates is to consider the varying perspectives that social groups have taken in them. It is safe to say that society is divided over most of these ethical issues, with one side in support of the benefits of technology, and others believing that the risk is not mitigated with respect to the technological benefit. In the case of abortion, for instance, the activists are poles apart, and the debate has continued with so little philosophical or ideological ground gained by either side that it almost seems natural that such a social division exists. Biotechnology is like any one of these groups. It has chosen the side that it supports, and it will continue to act consistently with that viewpoint. The people who make decisions at these companies believe that the potential benefits of the technology outweigh the potential risks, although they may not have acknowledged certain biases that afford this world view. Since social lines have been drawn, and they are not disappearing anytime soon, the responsibility to act bioethically lies with those on the side of progressing technology, and private industry will have to own up to this fact.

2

GENETICALLY MODIFIED FOODS

> In the last resort, the question to be asked here is not how much *man* is still able to do—there one may be sanguine for the Promethean potential—but how much of it *nature* can stand.
> —Hans Jonas, *The Imperative of Responsibility*

I. Executive Summary

Genetically modified foods represent one of the most controversial biotechnologies in recent history. Bringing hope to all types of groups, ranging from multinational corporations to developing nations, the science has simultaneously become their scourge. This chapter examines both the benefits and the risks behind this technology as it particularly relates to agriculture. After the opening examination, this chapter offers recommendations to those companies who practice this research to maximize the benefits while minimizing risks.

The Science

Inserting a foreign gene or genes into a plant to alter that plant's phenotype produces genetically modified crops. To accomplish this task, one of two methods is used. In the first case, modification occurs via a bacterium that usually kills wounded plants by inserting a DNA sequence that nurtures the bacteria at the expense of the plant's life. If the "lethal" gene from the bacterium's DNA sequence is removed and replaced with another, known as a transgene, the bacterium will insert the transgene into the host plant's genome. The other method shoots DNA-coated pellets into a plant cell's nucleus, causing the genes on the pellet to be incorporated into the plant's genome.

The Benefits

This technology has a substantial range of benefits. To aid the ecosystem, plants can be created to produce pesticides or resist herbicides. This decreases the use of chemicals in the environment, and eases the labor and costs associated with farming. Aside from these beneficial traits, health benefits can also be conferred by genetically modified foods. Plants can be engineered to lower their fat content or that of their derived oils, and they can produce vaccines and other therapies as well. More resilient species may also aid underdeveloped countries to feed their starving populations and improve their farming efforts.

The Issues

While there are many benefits, they do not come without controversy. The greatest reservations center on safety; concerned groups contend that altering the genetic basis of food puts those who eat them at great risk for unexpected side effects. At the same time, many believe that corporate interests force these technologies on society without the latter having a choice in the matter. Multinational corporations may be doing more than just pushing the technology on people, they may be manipulating plants in such a way to create a farming or national dependency on the firms that create them. Inequities, both global and domestic, can be furthered among the underprivileged, while many become suspicious of the scientists and executives who insist on developing this technology.

The Industry

To place the issues in context, the recent events at Monsanto and Aventis are examined. In the case of both of these corporations, their almost blind faith in the technology met with strong opposition. Placing too much stock in intermediary stakeholders, these corporations made the mistake of believing in their technical expertise and corporate strength strongly enough to ignore the criticism of those groups representing the end-consumers. Paying dearly for this oversight, they have now changed their attitudes and corporate agenda to accommodate the concerns of their critics.

Recommendations

To address these issues, while maximizing the benefits of the technology, recommendations are offered to help guide industry. Depending on the level of commitment that a corporation is willing to make, it may choose which, if any, of these recommendations it believes that it should adopt. For instance, labeling can be included with the foods as a means of informing consumers, while advisory boards can be formed to better analyze relevant issues.

Mastering agriculture has always required substantial effort, but success offered the potential of great societal accomplishment. Indeed, from its earliest inception, farming aided humankind's transition away from nomadic culture. With the promise of a harvest, and eventually the means to store it for use during harsh seasons, the shift to surviving within a single locale became less intimidating.

Farming, however, is not as simple as sowing seeds and waiting for the spring to usher in the ensuing bounty. Furthermore, distributing the food to the public is another difficult step, which depends on the acumen of the farmer and affiliates in the complex agricultural process. Farming requires hard work, understanding environmental patterns, complex machinery, storage silos, and experience with agricultural biotechnology, not to mention a means of protecting the plants from pests and herbicides.

This list was not always so comprehensive. Agriculture has evolved quite substantially over the ages. In many ways, the evolution of farming practices traces one of the longest historical tales regarding biotechnology that can be told. To hold a single ear of maize in one's hand is to look at the considerable progress realized by just a general understanding of botany. "Maize evolved from teosinte some 5,000–10,000 years ago, when hunter-gatherers in south-western Mexico started to select seeds of the 'best' teosintes they found in the wild and plant them deliberately. Their patient selection eventually transformed a straggly plant bearing its seeds in loose tassels on long side branches into the familiar maize plant, with seeds carried in handy compact cobs on very short side shoots."[1]

In the above example, human needs led to an innovation in crop technology: crops with higher yields and larger fruit were selected and propagated. Using agriculture as a tool to meet society's needs has a long and rich history, and in recent years, the potential has increased extraordinarily. In the past, crossbreeding techniques and cultivar selection had been used to create robust crops. However, new genetic technologies have expanded the type of traits that one might expect of a plant. Genetic modification (GM) alters the genetic structure of the plants by inserting a gene or genes from foreign species, which adds traits that were heretofore foreign to any plant, let alone the specific recipient.

The hope of this technology spans a wide breadth. From including vaccines to increasing the nutritional value of common food, the potential of "agriceuticals," as this technology is often called, has escaped neither the pharmaceutical industry nor the scientific community. Indeed, one scientist saw this application as a great solution to many of the world's most devastating problems. "Like a latter-day Johnny Appleseed, Dr. Ingo Potrykus, the German inventor of 'Golden Rice,' would like to send his seeds to poor people around the world at no charge."[2] Dr. Potrykus' "Golden Rice" contains a daffodil gene that creates beta-carotene, a necessary precursor to the metabolic process that yields vitamin A. This discovery could lead to supplementing the diets of many individuals in developing countries with the new breed of rice, particularly in those areas where the World Health Organization estimates that roughly 124 million children suffer grave effects due to

---

[1]Lawrence, Eleanor. The Maize Maze. *Nature Science Updates*. March 18, 2000. *http://www.nature.com*.
[2]Christensen, Jon. Golden Rice in a Grenade-Proof Greenhouse. *New York Times*. November 21, 2000. *http://nytimes.com*.

this deficiency.[3] The promise, however, of this technology has met with great scrutiny by protestors from many different groups. For instance, Greenpeace has attacked genetically modified plants ever since the technology first gained wide use; they also used Golden Rice as a specific example of how tinkering with genetic codes is not only dangerous in the worst-case scenario, but also useless in the best case. Citing the data offered by Syngenta, the company marketing Golden Rice, Von Hernandez, a Greenpeace spokesperson, emphasizes the fact that the average adult would need to consume up to ten times the normal amount of rice eaten in a meal to realize any nutritional benefit: "It is clear from these calculations that the GE [genetic engineering] industry is making false promises about 'Golden Rice.' It is nonsense to think anyone would or could eat this much rice, and there is still no proof that it can provide any significant vitamin benefits anyway."[4]

Greenpeace is not the sole critic, although it may be the most vocal. Many other groups have arisen to foster consumer opposition to these technologies. In Europe, public outcries that have led to many countries banning the import or export of such crops now resonate socially and economically as far away as Japan and the United States. Recently in the United States, Aventis CropScience recalled its GM product StarLink Corn, which accidentally tainted products ranging from cereal to taco shells, when it was reported that the genetically modified corn had found its way into human products rather than the animal feed for which it was intended. The recall is estimated to cost Aventis and the producers of the recalled products at least $100 million each, and it may, after all is said and done, add up to over $1 billion total. While there is evidence that StarLink Corn is not allergenic,[5] the EPA had not approved it for human consumption. The public unease over GM food was intensified by this error, resulting in lawsuits, moratoriums on import by many countries, and managerial restructuring at Aventis. It seems that the end consumer, not the farmers who buy seed, are affecting the biotech industry directly, despite their indirect reception of these products.

---

[3]Ibid.

[4]*http://www.biotech-info.net/asian controversy.html*

[5]Metcalfe, D. D. et al. Assessment of the Allergenic Potential of Foods Derived from Genetically Engineered Crop Plants. *Critical Reviews in Food Science and Nutrition, 36*(s), S165–S1866, 1996.

II. The Science

Genetic modification may not seem new to the world of agriculture, since the most primitive farmers selected seeds from plants that exhibited desirable traits to control botanical genetics. Indeed the promise of the Green Revolution, a transformation in crop characteristics that improved crop yield and endurance, emerged directly from crop hybridization experimentation. Beginning his work in the mid 1940s, Norman Borlaug crossbred many strains of wheat, eventually developing a hybrid variety popularly known as dwarf wheat. After careful selection and breeding, a variant was developed to produce higher yields, require fewer pesticides, and have a shorter stalk than its counterparts, which allowed it to bear more weight without toppling over. In 1970, Borlaug won the Nobel Peace Prize for his contribution toward the problem of world hunger; indeed, "the form of agriculture that Borlaug preaches may have prevented a billion deaths."[6]

Borlaug's accomplishment mirrors Potrykus' in three ways: both were intended to relieve the plight of those most effected by world hunger, both worked toward their goals by manipulating plant genetics, and both would spark the ire of environmentalist groups. In Borlaug's case, "by the 1980s finding fault with high-yield agriculture had become fashionable. Environmentalists began to tell the Ford and Rockefeller Foundations [Borlaug's sponsors] and Western governments that high-yield techniques would despoil the developing world."[7]

Despite continuing opposition, considerable though it is, genetic experimentation on agriculture will continue. These humanitarian roots have grown to bear scientific, social, and profitable fruits that have captured the attention of scientists, corporations, and biotechnology. New genetic technologies offer the means of improving known crops, as Norman Borlaug has, as well as creating completely new types, as Ingo Potrykus has done.

Although most crops can be classified as genetically modified, those that are the subject of this debate are more accurately termed transgenic. It is im-

---

[6]Easterbrook, Greg. Forgotten Benefactor of Humanity. *The Atlantic Monthly,* 279(1), 75, January, 1997.
[7]Ibid.

portant to understand this distinction because there is a significant difference between hybrid/crossbred crops and transgenic crops. The former have had their genome modified by breeding practices that not only have existed for years, but also have been used just as frequently in animals. Individual plants that have exhibited certain desirable traits were crossed to produce offspring that were artificially selected for specific characteristics, a breeding pattern that is performed and tested over the normal lifecycle of the plant. Whether the traits are manifest in the expected progeny will not be known until the crops reach the stage of development at which that attribute can be detected.

Transgenic modifications, the true focus of this chapter, refer to the insertion of a gene from a foreign body, either a plant or animal, into another organism. In this case, the new plant may acquire a completely new feature due to its modified genome. When a plant has a new gene inserted into one of its chromosomes, it is said to be transformed. Having created the new strain, the next challenge is to breed the plants; however, before that can be done, the successfully transformed plants must be distinguished from the untransformed varieties by a process known as cell selection. Cell selection is as important as genetic modification because it not only weeds out the undesired types; it is also critical to the cloning or regeneration process, which allows for the mass production of cultivars.

Before addressing the specifics of cell selection, it is important to discuss the transgenic process. After choosing the plant that will receive a new gene, known as a transgene, the trait to be transferred and its genetic source require identification. For instance, in the case of StarLink Corn, the goal was to transfer a gene that would obviate the need for pesticides. To do so, scientists needed to find a natural genetic source of a protein that would either kill or deter the corn borer, the pest that is the most threatening to corn. Researchers chose a soil bacterium as their best candidate, *Bacillus thuringiensis (Bt)*, whose toxic proteins (toxic to insects, that is), insecticidal crystal proteins, or delta endotoxins (Δtx) are *already commonly used as a pesticide*. The specific protein of interest, Cry 9c, and its corresponding gene were then isolated. Following the isolation, enzymes cut the gene from its flanking DNA. Purified to remove all of the unwanted *Bt* DNA, the specific gene was then fused with a native sequence of DNA that allows it to be both incorporated into the plant genome and recognized by the molecular machinery that transcribes and translates a gene into a protein (Figure 2-1).

Inserting this transgene into a plant cell is the next step in the process, and there are two ways to do this. One method involves *Agrobacterium tumifaciens,* and the other uses a device known as a gene gun. *Agrobacterium* is a bacterium that attacks wounded plants through an abrasion on the plant's surface. Upon entering the lesion, the bacterium is activated by the plant's natural healing process. While the plant is trying to close its wound, the *Agrobacterium* initiates a process where it cuts its own DNA and surrounds it with proteins that carry it to a cell nucleus of its host. In the nucleus, the foreign DNA is incorporated into the plant's genome, and the plant's own replication process will henceforth include producing nutrients to support the invading *Agrobacterium*. Through genetic engineering techniques, scientists have found a way to isolate the transported DNA and replace it with other sequences. Thus, the altered *Agrobacterium* is no longer toxic, and it can be used as a delivery system for any chosen transgene.

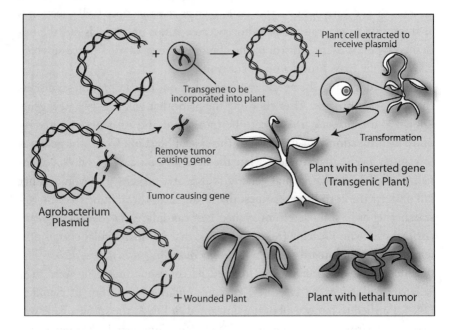

Figure 2-1   *Agrobacterium* transgene insertion.

The gene gun, although possessing a catchier name and very effective, is less elegant. Technically known as the biolistic method, this seemingly brute force approach requires quite a bit of finesse. As the name implies, this procedure works by "shooting" DNA into a plant cell. Gold or tungsten pellets are coated with a specific gene sequence, and the pellets are then shot into the cell, powered by gunpowder or helium bursts. The trickiest part is calculating the amount of force used—enough force must be used to penetrate the cell membrane, but it should be gentle enough to keep the DNA pellet from exiting the other side. Once inside the cell, the details of its incorporation and the efficacy are somewhat of a mystery, but the assumption is that the DNA dissociates from the pellet and the cell's organelles incorporate the DNA into the plant genome (Figure 2-2).

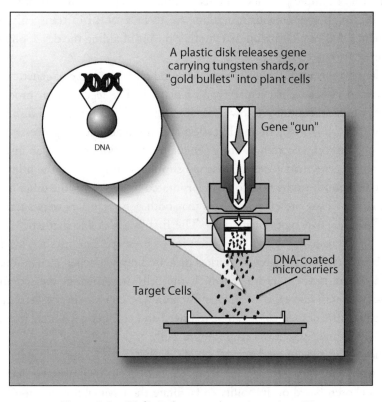

Figure 2-2   Biolistic (gene gun) transgene insertion.

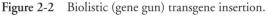

Although there are many types of technologies that aid the transformation process, perhaps the most important complimentary technology is a means of cloning plants to propagate specific types. The cloning methodology is an adaptation of somaclonal variation (SCV), known as cell selection. Because any given plant cell is totipotent—that is, it can grow into any cell produced by that species and can differentiate to form a distinct organism—a single set of cells can be cultured in a Petri dish to induce plant regeneration, or cloning. With the right mixture of nutrients, a plant tissue sample can grow into a new plant of seemingly identical genetic composition. However, not all of the regenerated plants are identical. Many will have slight genetic variations, or mutations, that manifest in physiological oddities. Although this may seem like a drawback, it offers a rapid way to induce genetic variations that might ultimately prove beneficial to agriculture. Because of its multiple products, the technique was termed somaclonal variation. Although spontaneous mutations are to be avoided in transgenic experiments, this peculiarity offered another means of aiding the development of these cultivars.

Scientists speculated that the method could take advantage of genetic differences to choose a particular type of plant in the regeneration process, which, when successful, is known as cell selection. Within the Petri dish full of plants, both the genetic modification and the regeneration processes are used to aid in cell selection. During genetic modification, rather than inserting a single gene, two genes are inserted simultaneously. While the primary gene corresponds to the trait to be incorporated by the plant, the other gene is a marker whose presence indicates that both gene transfers were successful. There are two types of markers. The first is a scorable marker, which functions by responding to a simple test for its presence. For instance, a gene may be inserted that creates a colorless protein; however, when a chemical is added to the nutrient broth in which the cells are growing, the scorable marker's protein may react to change its color. The colonies in the Petri dish that exhibit the desired color can be removed and cultured separately to produce larger quantities of the crop (Figure 2-3).

The second type of marker, a selectable marker, takes greater advantage of the regeneration process than the scoreable marker. In this case, the marker gene is chosen based on its ability to facilitate the regeneration of modified

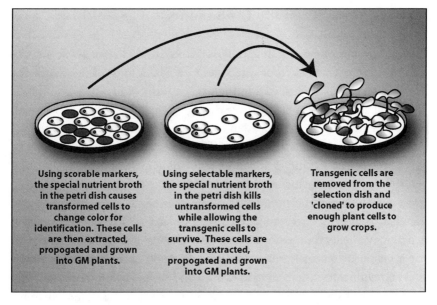

Using scorable markers, the special nutrient broth in the petri dish causes transformed cells to change color for identification. These cells are then extracted, propogated and grown into GM plants.

Using selectable markers, the special nutrient broth in the petri dish kills untransformed cells while allowing the transgenic cells to survive. These cells are then extracted, propogated and grown into GM plants.

Transgenic cells are removed from the selection dish and 'cloned' to produce enough plant cells to grow crops.

Figure 2-3   Scorable and selectable markers.

cells in conditions that impede the growth of unmodified cells. The most common process involves inserting an antibiotic resistance gene into the genetically modified plant. For instance, the *npt-II* gene confers resistance to the antibiotic kanamycin. By infusing the nutrient broth with kanamycin, untransformed cells will die and modified cells will enjoy a growth advantage. To simplify the process the selectable marker can be the gene of interest, particularly when herbicide resistance is the goal of the modification; by including the herbicide in the growth medium, the resistant strain is given the selective advantage.

Taking advantage of the many tools offered by genetics and molecular biology, genetic modification represents one example of taking these tools out of the lab and placing them into a production line. Companies have been quick to lead this transition, and they do so with the belief that the products will be both lucrative and socially valuable. Exploring the truth behind these claims, the following section may also hint that the advantages are almost as obvious as the reservations surrounding the technology.

III. The Benefits

The applications of genetically modified plants are as diverse as the advantages they offer. There are four categories into which most of these benefits fit, all of which will be revisited in the next section because they also frame the ethical debate. Despite the diverse range of genes that can be inserted into the plant genome, most are aimed at meeting one or more of the following goals: aiding the environment, increasing nutrition, meeting the dietary needs of our growing population, and aiding medicine and healthcare.

Although environmental groups represent a large proportion of the opposition to genetic modification, many of the developments in this field were intended to alleviate environmental stress. Indeed, in one of the environmental movement's foundational texts, *Silent Spring,* Rachel Carson explains that a major factor challenging a sustainable world is the overuse of pesticides and herbicides.[8] According to Carson and many environmentalists, the disproportionate application of these compounds threatens the environment, particularly biodiversity—an effect that deserves attention for its impact on the future of the human race as much as any other species. Industrial farming is the main source of the introduction of these chemicals into the ecosphere, and no matter where else changes are applied, any effort to curb the pattern *must* be made in this sector. On average, up to six different toxins aimed at destroying insects or weeds that threaten a crop's success are applied per crop per harvest season. Finding a means of obviating the "excessive" use of these herbicides and pesticides might be seen as an ecological imperative as much as a commercial stratagem.

Although the motivation for developing genetic modification for environmental purposes may be attributed as much to the ecological benefits as to its appeal to farmers, the latter deserves elaboration. Pesticides and herbicides are considerably expensive because their effectiveness renders them a competitive necessity. However, continually dousing an area with the same types of compounds is limited in effectiveness because the large volumes applied often aid their targets in selecting for resistant strains. In the interim,

---

[8]Carson, Rachel. *Silent Spring.* Boston: Houghton Mifflin, 1962

the population of natural predators for the pests might decline due to the widespread drop off of prey. Upon the resurgence of resistant insects, neither the pesticides nor the natural predators can protect the farmer's investment of time, effort, and capital in these crops. Although pesticide and herbicide resistance remains a consistent threat with any type of biologically based deterrent, a more precise method like genetically modifying plants to express toxins specific to pests and resistant to herbicides would delay or even interrupt this cycle. The result would obviate the need for applying pesticides that can spread beyond the farming environment; furthermore, the herbicide resistance can be engineered toward withstanding one specific type of herbicide, thus reducing the number of different chemicals applied. In the long run, labor also decreases, as better management of pests reduces the amount of tilling and plowing, which in turn preserves topsoil and reduces the runoff of sediment and fertilizer that compromises local stream and river ecosystems. The latter two benefits exhibit how economic benefit to the farmer and the potential contribution to environmental subsistence can be accomplished simultaneously.

Choosing a transgene to meet environmental needs, however, offers a bit of a challenge. The criteria for being environmentally friendly have to be balanced with the concept of efficacy. For instance, expressing a pesticide-like toxin cannot be accomplished without a natural source for the gene and its corresponding protein. Luckily, industrial farmers have been using a pesticide derived from an organism for years—the soil bacterium *Bacillus thuringiensis* naturally expresses a family of proteins (over 100) that are insecticides. These proteins have long been a standard part of a farmer's arsenal against pests, which could not have become customary in the United States without the approval of the Environmental Protection Agency. Given their proven track record, they seemed the most logical starting point. Isolating the gene that corresponds to a plant's defense against its most common predators, or a set of genes for predators, and placing it within the plant's genome is made that much simpler by the already common use of *Bt* toxins.

In a more general sense, this technology is more than a means of developing crops for consumption and commercial use. The very real threats of declining biodiversity and deforestation can be mitigated by advances in genetic modification. Endangered plant species can be preserved and propogated with this technology. Also, species of plants that are declining in upset ecosystems can be adapted to face the adversarial environment until the balance is regained. Finally, as the technology advances, plants may grow in higher yields to replenish barren areas faster than by conventional methods.

While pesticide- and herbicide-resistant GM plants are in wide use, those that offer increased nutritional value will be the second generation of transgenic plants. Often included under the term "quality traits," these improvements are noticeable changes that directly benefit the consumer, rather than the farmer and the environment. While current GM improvements may lead to lower prices, which is yet to be realized, quality traits may be altered to increase the vitamin content of a grain or decrease the saturated fat in a seed. For instance, the Golden Rice mentioned previously is exactly this type of alteration, which will become more common as the technology is standardized. Although Golden Rice does not currently offer an adequate daily source of Vitamin A, further experimentation will lead to second and third generations that will meet if not exceed requirements. Although Golden Rice is aimed at alleviating nutritional deficiencies in developing countries due to a lack of food, similar advances are underway to relieve the plight of first-world countries that live with excess. The rise of heart disease, obesity, and other health conditions in the United States are often attributed to poor dietary habits, and they are often countered through dietary measures. Cancer and other illnesses may also include dietary restrictions as part of a treatment regimen. To meet these requirements, genetic modifications might be made to oil seed, commonly used to make cooking oils. The oil seed may be altered to decrease its saturated fat content, and it can further be modified to include more vitamins, antioxidants, or other nutritional elements.

These benefits represent the greatest promise of the industry, not because the others are less important, but rather because these applications directly influence the general populace. So many of the other uses of GM technology are biotechnological marvels, but they do not resonate with the most common denominator, individual consumers. We are all individual consumers, even the CEOs of biotech firms; our concerns are those that matter most because when we change the world through science, no matter which institutions are most profoundly affected, it is a collection of individuals who will support or denounce that change. As a reader of this book, you may be changing your mind about the technology, or more deeply entrenching yourself in your previous conceptions, but your attitude will decide whether you accept or deny the produce in your grocery store. Furthermore, if biotechnology can get into the produce section of the local "Stop N' Save," then it can go anywhere, physically and metaphorically. This is the revelation that stands in the way of private industry, and if biotechnology's pervasiveness confers risk without benefit, why would anyone, even the Board of Directors of a biotech company, support it? Although this concept is explored in more detail in both the issues and industry sections of this chapter, its conceptual importance deserves mention here as well.

When one considers the Green Revolution, its goal to end world hunger through the introduction of better botanical technology was ambitious and worthy of the award for peace that the Nobel committee conferred upon Norman Borlaug for his contribution. By breeding hybrid plants—a genetic modification because the process manipulates the genetic composition of a species, but not a transgenic modification because it does not use recombinant DNA technologies—staple crops in developing as well as developed countries can produce higher yields and resist more environmental challenges. The further goal was to let these seeds alleviate economic strife while relieving hunger in developing countries by helping the farmers gain a higher, more stable return on their planting efforts. While the Green Revolution has not fully eliminated these problems, it has done a tremendous job in bringing the crisis closer to an end. GM foods offer another leap towards this goal. As more advances are made in understanding plant genetics, the means of improving and specifically tailoring each crop to its environment will emerge. Furthermore, the difficulty of having to buy hybrid seeds every season because the full-grown plants that grow from them do not reproduce well, if at all, will be obviated. The result, at its most ambitious (barring political interferences between countries), may lead to self-sufficient farmers, and at its least, it will add another weapon in the arsenal in the war against hunger.

Another type of "quality trait," which is mentioned last in this portion of the discussion because the technology is furthest away from implementation, is the introduction of therapeutics into GM food. Biotechnology has a substantial history of introducing novel therapeutics to healthcare. Traditionally, all drugs, old and new, have been delivered in pills or syringes, but GM technology offers a means of changing that. Some day, rather than giving children a series of shots for vaccinations, the treatment will be administered through their broccoli or carrots. Although it may be hard to make children eat their vegetables, it is still easier than giving them shots. The greatest benefit will be realized in developing countries where hunger and sickness can be addressed simultaneously. The further expense of syringes and other intermediary technologies will be removed, and many diseases that have been eradicated in more developed areas will stand closer to ending their plague in less fortunate populations.

These represent a sampling of the immediate projects for which biotech firms have been trying to use GM technology. A host of other applications are on the roster for the future. Ranging from farming advances to materials development, the world may see such innovations as drought- and frost-resistant produce. As a greater understanding of plastics and fuels emerge, commodity chemicals may be grown in plants, introducing organic plastics and replacing standard fossil fuels. The limits of the technology may rest solely on the shoulders of the imagination of humanity—a prospect as exciting as it is foreboding.

IV. The Issues

GM foods present a very interesting set of issues that frame the opposition; objections to the technology read as a point-for-point refutation of the benefits. More interestingly, both sides of the debate rely on scientific evidence, and in some cases the same studies support both factions. Ironically, a field that prides itself on approximating truth through empirical analysis is providing the information whose varied interpretations set the grounds for dispute. Scholars in the discipline of science studies have long understood the fine yet weighted difference between *data* and the *knowledge derived from data*. Whereas the former is a collection of "facts," the latter is an interpretation of those facts within a system of thought. Ideally, however, the careful, rational analysis of the information will lead to a better understanding of its value and place in society, although science and society are very closely related, with every indication pointing toward greater confluence between the two before any signs of divergence surface. Science plays an important role in the everyday functioning of society, particularly as we become more dependent on technology; if a company wishes to aid society, it should understand the demands and reservations of these stakeholders, as defined in its broadest sense. In a speech she delivered in 1952 during her acceptance of the National Book Award for *The Sea Around Us*, Rachel Carson made this point:

> We live in a scientific age, yet we assume that knowledge of science is the prerogative of only a small number of human beings, isolated priestlike in their laboratories. This is not true. The materials of science are the materials of life itself. Science is part of the reality of living; it is the way, the how and the why for everything in our experience.

Reflecting the topics introduced in the Benefits section above, the first area of critique in this discussion regards environmental issues. Most adversaries of GM food cite the technology's novelty as the primary problem; more precisely, altered crops are feared as too new to be placed in the precious arenas of the environment and the dinner table. Continuing the argument, the potential risks involved necessitate that society reject them—placing these new, unnatural organisms in the world may result in catastrophic damage. More concretely, concerns center on upsetting the ecological balance, as well as the further damage that humans might levy upon the environment due to changes in agriculture spawned by GM plants.

Environmental issues deal with the effects on animals, plants, and microorganisms alike. In the case of animals, the worry rests on the damaging effect that toxins such as *Bt* might have on rare, endangered and even healthy populations of wildlife. Perhaps the most famous example is the tale of the monarch butterfly's encounter with a GM crop. A symbol of vulnerability and beauty, this creature became the subject of one of the most heated debates over science in the past decade. In early 1999, entomologists published a widely quoted paper in the respected science journal *Nature*. In this paper, the Cornell-based group claimed that *Bt* pollen was deadly to monarch larvae.[9] Both the press and opponents to GM food jumped on the story; it was good copy and among the strongest evidence against the technology. Compromised, this high-profile species' plight made the threat of GM foods seem more real. Almost immediately following the publication of this paper, a series of scientists criticized it as preliminary and inconclusive.[10] When a more comprehensive study was performed, notably by a consortium funded by a series of agricultural biotechnology companies, these results and others showed that the *Bt* toxin was *not* a threat to the monarch and other types of butterflies.[11] However, the point had been made that it is not unreasonable to assume that genetic modification may affect species beyond the insecticidal target.

The concern over biodiversity, as the furor over the monarch illustrates, is an important topic to the public. As the refutations have indicated, the threat is minimal, but a point cannot be ignored: *Bt* toxins are used to kill insects that prey upon crops. Many insects, like the corn borer for instance, are targeted for extermination because they attack crops. There may be an overlap in pesticidal efficacy between target and nontarget species, but the more important fact is that pesticides are in use and will always be in use, whether transgenic or chemical. Chosen for their properties, *Bt* or any other commercially available pesticide has been approved by the EPA for use on crops. Indeed, the *Bt* transgene that is inserted is chosen by means similar to the design of a chemical pesticide. Banning GM foods avoids the true issue, which is the threat to biodiversity via insecticides. This explanation is not meant to belittle the issue of the environmental concerns that GM crops represent; rather, it is meant to clarify the difference between an argument and choosing one scapegoat for its ire. Threats to biodiversity distress the public, and if one technology raises the issue to a more noticeable dialogue despite misdirected antagonism, the public good is still served by the recognition; however, it ought to be clarified, rather than reinforced in error.

---

[9]Losey, J. et al. Transgenic Pollen Harms Monarch Larvae. *Nature*, May 20, *399*, 214, 1999.

[10]Hodgson, J. *Nat. Biotechnol.*, *17*, 627, 1999; Beringer, J. E. *Nature (London) 399*, 405, 1999; Shelton, A. M. & Roush, R. T. *Nat. Biotechnol. 17*, 832, 1999.

[11]Kendall, Peter. Monarch Butterfly So Far Not Imperiled. *Chicago Tribune*. November 2, 1999.

Another point that falls into a similar category as the mistaken interpretation of GM pesticides is the emergence of resistant pests. Critics of the technology claim that transgenic plants that release insecticide are aiding in the selection of resistant strains of the insects that were targeted by the transgene. Already, the argument against this point is probably forming in the reader's mind, following the logic in the previous paragraph: the risk involved has little to do with the means of introducing the toxin, and everything to do with the use of pesticides. It does not matter from where the pesticide comes, just that its presence might select for more resilient bugs.

Upsetting biodiversity has a broad affect on an ecosystem's cycle of life; like dominos, compromising one species may harm others that depend on it. Although animals like the monarch butterfly are the subject of concern, so are plants. Although the issues surrounding animals focus on extinguishing species, the debate regarding plants center on creating new varieties. Often termed "superweeds," the misgivings about the new types of plants are in response to a process known as "gene jumping" or "gene flow." In this process, a gene from one species "jumps" into another; essentially, a transgene might be transferred to a species of weed by the GM plant's natural reproductive process of releasing pollen into the atmosphere. If this transgene corresponds to herbicide tolerance, then the resulting weed may be resistant to the herbicide. A species of superweeds may take over the surrounding, wild environment because the usual herbicide will no longer kill the weeds.

Gene jumping, as described above, may seem a serious threat, but there are strong, effective arguments from the GM supporters that refute it. Perhaps the most obvious reason is displayed in the lapse in logic of the above example. The case stated that herbicide resistance would create a breed of superweeds in the wild that would resist the chemicals; however, herbicides are not sprayed willy nilly into the environment. Because there is no selective advantage to herbicide-resistant weeds in the environment, there is little reason to worry. Indeed, a recent study spanning 10 years shows that the problem is minimal at best.[12] According to Michael J. Crowley of Britain's Imperial College, the paper's author, "In no case was the GM crop more invasive

---

[12]Crawly, M. J. et al. Biotechnology: Transgenic Crops in a Natural Habitat. *Nature, 409*, 682–683, 2001.

than its conventional counterpart. The risks with these crops and these constructs truly are negligible."[13] Although this may not seem a major issue because the most popular concerns proceed on shaky grounds, it does deserve attention due to the types of genes that are being inserted. Because there is the potential for gene jumping (a number of cases have occurred naturally between plants of high and low degrees of genetic homology), careful thought must be given to the type of genes considered for insertion, as no one wants to release traits that would be dangerous if adopted by random species in the wild.

For quite some time, the concern surrounding gene flow was limited to plants of similar genetic composition, but when evidence showed that two species with different numbers of chromosomes can swap genes, the stakes were raised a bit higher. According to Allison Snow, a botanist at Ohio State University, "What really shocks me as a biologist is that you have two species with different numbers of chromosomes hybridizing. Goatgrass has 28 chromosomes and wheat has 42, but they can cross."[14] Even more unsettling is the prospect of "horizontal" gene flow across species—the transfer of genetic properties through nonreproductive methods. Recently, an article in *The Observer* stated: "A four-year study by Professor Hans-Hinrich Kaatz, a respected German zoologist, found that the alien gene used to modify oilseed rape had transferred to bacteria living inside the guts of honey bees."[15] Kaatz, reluctant to speak prior to publication given the history of premature publishing of anti-GM data did explain: "It is true, I have found the herbicide-resistant genes in the rapeseed transferred across to the bacteria and yeast inside the intestines of young bees. This happened rarely, but it did happen."[16] Creating modified bacteria or viruses may be a considerable threat to human health and other organisms should transgenic species grow unchecked; thus, genes should be chosen based on their safety profile, which should, in turn, be weighed against the potential for horizontal gene flow.

---

[13]*http://www.cnn.com/2001/TECH/science/02/07/modified.crops/*. February 7, 2001.
[14]Mann, Charles C., Biotech Goes Wild. *MIT Technology Review,* July/August 40, 1999.
[15]Barnett, Antony, GM Foods Jump the Species Barrier. Sunday, May 28, 2000. *http://www.observer.co.uk/uk_news/story/0,6903,319418,00.html.*
[16]Ibid.

The critique against the nutritional value of GM food should not be aimed at the technology because it relies on the false assumption that the science behind the technology is complete. Using Golden Rice as the exemplary case, most challengers take the position that the technology cannot deliver what it promises. A trace amount of Vitamin A in Golden Rice does not offer any real advantage to the consumer; thus, the opponents contend, a novel technology is introduced into society at great risk for no benefit. It also seems that this technology, aimed at developing countries, is being pushed on the less fortunate in lieu of interventions that would substantially help an undernourished population. Although releasing this type of crop in a country where it may show no benefit makes little sense, it becomes difficult to use this reasoning as the sole argument for halting research. No one can expect that every technology will work immediately, and further experimentation is necessary to provide vegetables with advantages like more calcium or less fat. More work must be done before the goal is met to grow produce with substantially increased vitamin content. There is little doubt that such advances will occur given enough time and effort. However, if a company states that its goal is to immediately work toward the relief of malnutrition and if no effort is placed on using its conventional technologies to do so, then it is not unreasonable to assume that the interest is in distributing technology, rather than helping a hunger-stricken population.

The more salient analysis of this topic does not concern nutrition, but the place of multinational biotechnology companies in the developing world, as well as its effect on poor farmers domestically. Safety and industry come into conflict here because technological efforts to mitigate risk can also be seen as contributing to monopolistic business practices. Most pointedly, regarding questions of gene flow, the obvious solution would seem to be genetically engineering a means of preventing reproduction. A genetic modification can be introduced into plants that prevent them from going through a full life cycle, thus removing the plant's ability to release its genetic material into the environment. The technology, commonly known as Terminator technology, uses a promoter region (a DNA sequence that signals a cell to start making a protein from the gene to which it is adjacent) and a gene that codes for a protein that halts seed production. In this case, three types of DNA sequences are integrated into the plant genome: promoter from the cotton

Late Embryogenesis Abundant (*LEA*) gene, a specific DNA sequence that separates *LEA* from the last inserted portion, which is the Ribosome Inhibitor Protein (*RIP*) gene. During the process of development, an enzyme that naturally occurs in the plant removes the sequence that separates *LEA* and *RIP*. When the plant reaches the stage in its development that immediately precedes creating seeds, another of its naturally occurring enzymes recognizes the *LEA* sequence and initiates the molecular machinery to create the *RIP* protein, which kills the seeds. Ingeniously, the *LEA* sequence was chosen because it is only recognized in the seed development process; it has no function in the rest of the plant. The plant functions normally, and only its seeds are affected by the genetic modification (Figure 2-4).

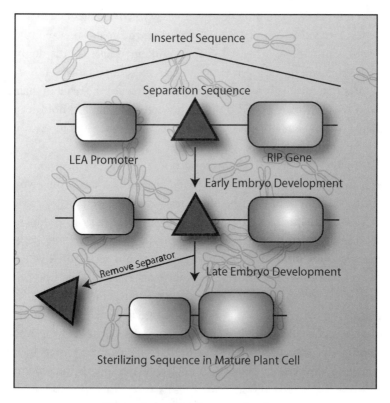

Figure 2-4   Terminator technology.

Although the technology solves one biological problem (the risk of spreading GM traits), it creates a socioeconomic one. This transgenic feature is also a method of copy protection, which means that any farmers who are dependent on these plants cannot reuse seed. To plant another season's harvest, the farmers must buy a new set of seeds. Domestically, this practice is fairly standard, and it protects the corporate investment in developing the cultivars, but it still limits smaller farmers and developing nations. In the case of the latter, those farmers and politicians who welcome these contributions by large corporations place themselves in a position of dependence on the companies. Socioeconomic–political dynamics can be substantially changed through the influence and interference of foreign interests. As important as agriculture is to any society, there is little that a hungry, malnourished nation can do but naturalize a monopoly if it has already taken root in the country's food production industry. As new generations of Terminator technology emerge, they will foster even greater dependence on a company's products; not only will biotechnology firms sell the seeds, they will have developed ways to control their activation, rather than deactivation, and they will regulate the sale of chemicals that direct this process. Although none of these power plays have been made, their threat requires assurances that corporations will not endanger foreign cultures.

The risk to human health may be the most alarming claim made against the creators of GM crops. All of the other concerns will affect the end consumer if they are realized, but they do so indirectly. Understandably, individuals who will eventually eat the food want to know if the inserted gene will harm them. Perhaps the most famous case involves the Brazil nut and soybeans. Brazil nuts are a source of methionine, an essential amino acid, which soybeans do not produce in high abundance. Although the high protein content of soybeans gives it a prominent place in animal and human diets, the absence of the amino acid is an area where diets could be supplemented through the technology. Scientists isolated the gene for methionine production from the Brazil nut and placed it into the soybean. The experiment was a success in that the modified soybean exhibited substantially higher levels of methionine; however, there was still more testing to be performed. Since Brazil nuts are allergenic to some individuals, the company developing the product, Pioneer Hi-bred, ordered a battery of tests to determine whether or not the transgenic soybean would elicit a reaction in someone allergic to Brazil nuts. The answer was yes. Although Pioneer immediately stopped the project, the threat that allergenic properties could be inserted alongside a transgene became a reality. Critics recognized that identifiable allergens could be detected through standard methods, but new allergens might be introduced because of the wide variety of traits and source organisms that can be exploited by the technology. These concerns have emerged partially because of GM foods, but also because of the advancing field of xenotransplantation (organ transplantation between species). Research in this area has explored why the human body rejects foreign elements, and in some cases a severe immune response may ensue after the new environment awakens viruses that are dormant in other species. While these are all possibilities, it must also be recognized that the early detection of the Brazil nut allergen in the soybean is an indicator of the success of the regulatory process—the product was never released, and there was no exposure outside of a controlled setting.

Another threat to human health deals with antibiotic resistance. An increasing number of bacteria are becoming resistant to antibiotic drugs; the overprescription of antibiotics caused this phenomenon.[17] Using selectable markers, antibiotic-resistant genes can be inserted with the transgene allowing GM plant cells to remain alive during the regeneration process, thus increasing the exposure of the drug to the environment. The justifiable fear is that, as more and more bacteria come in contact with the markers in the wild, the more likely it will be that these plants will help "naturally" select resistant strains for survival. A possible pandemic of untreatable bacterial infections might arise due to human intervention in nature.

V. Industry

Unlike many of the genetic technologies that have been introduced over the past few decades, GM foods fit within an already established industry: agri-

---

[17]Levy, S. B. *The Antibiotic Paradox: How Miracle Drugs Are Destroying the Miracle.* New York: Plenum, 1992.

culture. Although the technology's domain in the future will expand to include therapeutics and other types of applications, GM innovations currently focus on improving farming, thus placing it within a tradition that has already faced some substantial criticisms. Unfortunately for industry, and perhaps fortunately for society, a historical wariness has surrounded the industry and its regulations, which is not surprising since this business depends on consumers eating the products grown from corporate seeds. Food has a cultural value ranging from necessary, worldly sustenance to the sacred; thus, its most mundane place in society is not mundane enough to escape scrutiny. In a sense, the attempt to bring the technology from the lab to the dinner table was destined to be an uphill battle. Two general examples of the furor that frames the oppositional, historical precedent are the commercial introduction of hybrid plants and the regulatory failings regarding European livestock practices. These critiques set the foundation for the discussion that follows, which shows the negative societal image that Monsanto and Aventis had before them as they brought genetically modified foods to the world.

Many industries have had to counter the critique that the means of production, if put in use to serve the ends of corporations, might limit social freedom. This lofty expression calls names like Karl Marx to mind, who often seems too esoteric to have any practical meaning in the realm of business. However, by ignoring the critique because it has an aversion to scholarship, the industry assumes a position in which it may throw the baby out with the bath water. Furthermore, science and its relative objectivity is often thought to be beyond the reach of these types of criticisms, but understanding how the critique applies can help in clarifying the opposition's views with respect to multinational scientific corporations. Although many problems that escape the broader public's notice may seem irrelevant, the discerning eye of an insider might be a better gauge of problems to come. How genetics have been appropriated for industrial rather than social ends has been identified by a number of scholars, not least amongst them noted geneticist and philosopher Richard Lewontin. In his analysis in *Biology as Ideology: The Doctrine of DNA,* Lewontin acknowledges that the introduction of hybrid plants offered many benefits to society; however, the scientists and corporations behind the advances held the cards in determining which potential benefits would be investigated. Trying to decide what to pursue, scientists at Pioneer Hybrid Seed Company (now Pioneer Hi-bred) decided to manipulate hybrid genetics to create robust crops that could not reproduce equally robust offspring. First developed in 1924, the Pioneer Hybrid seeds, a predecessor to Terminator technology, were a means of copy protection. Aiding Pioneer were certain political appointments that placed hybrid seeds at the forefront of the agenda of the then Secretary of Agriculture, who coincidentally was the father of Pioneer Hybrid's CEO. Lewontin recognizes that the "invention of hybrid corn was, in fact, a deliberate use of the principles of genetics to create a copy-protected product. . . . [However] By the method of selection, plant breeders could, in fact, produce varieties of corn that yield quite as much as modern hybrids. The problem is that no commercial plant breeder will undertake such investigation and development because there is no money in it."[18] Agriculture, then, is one of the first industrial applications that enjoys the dubious distinction of exploiting genetics for corporate gain. Of course, it is generally understood that corporations have fiduciary obligations which have to be met, both in terms of law and business ethics; however, recognizing this distinction is critical to understanding the bioethical framework in which the industry can fit.

Of the types of responses that the Pioneer Hybrid Seed Company might elicit, two are most prominent. The first has to do with equating commercial success with social success; a product will not be commercially successful if it bears no benefit to its consumers. In this case, the only difference between traditional and hybrid varieties is that the latter are generally better than standard reusable seed, so the choice is left to the effected consumer— the farmer. Despite the utility, the question as to whether selling the seed creates an immediate dependence on the distributor does not vanish; with all of the other factors involved in commercial competition, political pressures not least among them, whether there is any choice in the matter is the adversarial question that will not go away. The second response builds directly on the first and is the issue explored more thoroughly here, the question of trust. Food is too important a factor in everyday life for anyone to idly accept a product that has some risk attached to it, and although this is-

---

[18]Lewontin, R. C. *Biology as Ideology: The Doctrine of DNA*. New York: HarperPerennial, 1991, pp. 55–56.

sue did not preoccupy the world upon the advent of hybrid technology, it is *the* domino that begins the tumble of GM technology. Compounding this point is the fact that for quite some time only fringe groups concerned themselves with the products being placed on grocery shelves, but a series of events mishandled by government agencies played into activists' hands to bring the issue to the world's stage, a tradition that will continually be observed as biotech becomes recognized for the global agent that it is. In a series of events in Europe that occurred without influence or interference from companies selling GM foods, the agencies that were responsible for ensuring the safety of food products systematically failed at their responsibilities. The outbreak of mad cow disease, scientifically known as bovine spongiform encephalopathy (BSE), proceeded to worsen due to the failure of regulatory bodies. Although the disease had been recognized as a problem as early as 1986 in England, the restrictions put in place to limit the spread of BSE were poorly enforced. In 1996, the scope of the problem increased when a new variant of the human neurodegenerative disorder Creutzfeldt–Jakob disease was discovered and linked to the spread of BSE. Mad cow disease had spread largely due to the lack of intervention on behalf of the Ministry of Agriculture, Fisheries and Food (MAFF), which did not enforce its regulations halting the reuse of slaughtered cattle as animal feed. The recycling of animals included the recycling of BSE until today when the practice is banned, but the damage is done. Now that public trust in the regulatory agencies has been lost, why would anyone rely on the word of the corporations on the other side of the regulatory fence? Also, to whom should corporations in the EU turn to validate the safety and benefit of their products in the eyes of their stakeholders? Perhaps jumping the gun, Monsanto answered this question upon its introduction of GM foods in exactly the wrong way.

More than any other company in recent history, Monsanto may have received the brunt of the public opposition to biotechnology. In 1997, reports of the company's transgenic soya beans mixing with unmodified beans emerged in Britain, and consumer outrage quickly followed as critics claimed the mix-up was an effort to force these products on the populace. Since that moment, GM foods have become the symbol of the perils of industrial biotechnology, and protestors were ready to respond on the worldwide stage. Activists from such high-profile organizations as Greenpeace have global networks that can organize and rally against any corporation almost anywhere in the world. The response to this type of attention by corporations is often to tighten its lips and follow its business strategy; however, there is little a firm can do against the information network built on the Internet. When Monsanto continued to extol the virtues of GM food, the public was deafened by voices of protest that painted an apocalyptic picture with GM crops as brushes. In the past, bits of information and misinformation might not find a nexus, but the Internet allowed many different opinions and much of the opposition to do two things: the first was to consolidate activists from all over the world, and the second was to impose a corporate "transparency" on Monsanto. Transparency, the opening up of operations to the public to allow assessments and judgments of the firm's functioning and strategic goals, is one of the few tools that activists accept as proof of a corporation's dedication to society. With the introduction of the Internet, it is almost useless to avoid transparency over a controversial technology when a critical group takes aim at it. By not inviting critique and understanding, the public has to rely on those few people who are talking about the issues, and if all of the declarations are negative, the opinions that form will reflect this.

Although there are different degrees of intensity in the United States versus the European Union, the furor is always negative, and will continue to be so. There are peculiarities about the mode of introduction for GM foods that elicit vocal responses from only the opposition. When Monsanto decided to release their first GM products, they made a bad mistake. Making perhaps *the* paradigmatic error in corporate biotechnology marketing, Monsanto thought of their product simply as a product. As such, GM foods were positioned by the same type of strategies used for all general products: choose the target market and get that market to buy as much of the product as possible. In this case, the target market was the farming industry, and this is the critical mistake that Monsanto made. They pushed the technology into the marketplace while identifying only a single stakeholder to whom they would market, the farmer. The first generation of products made farming easier; Roundup Ready soybeans, for instance, were resistant to Monsanto's Roundup herbicide, making spraying the crop simpler and cheaper.

More importantly, by creating a first generation of products that only benefited the grower, there were no recognizable advantages to the consumer; price, nutrition, and taste, noticeable to all, were not improved. Whether such advantages might help the environment was incidental and less important compared to the potential risk that might be levied by this new technology. That, however, did not stop Monsanto from trying to convince the world that the advantages of GM foods extended to everyone. This claim has led to protests that range from boisterous opposition at conferences such as BIO 2000 to destroying GM crops on private property.

It may not seem that creating a product that has tangible benefits could be a mistake, but the problem gets to the heart of the tension between biotechnology and society. When dealing with something as pervasive as food, a commodity whose necessity has no cultural or geographic bounds, a symbol is quickly created that represents the reach of biotechnology. Because this technology is built on genetics, a universal characteristic of all living organisms, nature and culture are seen through this example as being brought together in a new, necessarily troubling way—necessarily because such rapid changes that can alter social, political, and economic institutions deserve careful scrutiny. Although these issues may not reach the forefront of concern of every corporate biotechnology firm, it does not follow that the engineers behind the revolution should remain silent because they have a business or science degree instead of one in the social sciences or the humanities. Daniel Vasella, Chief Executive of Novartis, a competitor of Monsanto's in the GM food space, realizes this point: "We are not missionaries. . . . We sell things. No company can prosper by telling customers what is good for them. . . . This is not just about plants. It's about our myths, our history and culture. It's about what we put in our mouths and in our babies' mouths. When you go to somebody's house, they offer you food. That is a ritual of everyday life. What is more basic—and what could be more frightening— than playing with that? Of course it scares people. How could something this important not scare people?"[19] The world is changing rapidly, and more than any other time in history, humanity is the architect of this change. Sci-

---

[19]Specter, Michael. The Pharmageddon Riddle. *The New Yorker,* April 10, 2000, p. 60.

ence is a tool that has the ability to fundamentally alter our way of life, and, more importantly, our way of understanding how humans fit in the world. When a product becomes the archetype of this theme, all those who recognize the implications of biotechnology, regardless of the severity, will rally for or against the new emblem. Too bad for Monsanto that the symbol became GM foods.

To those sympathetic to biotechnology and the plight of Monsanto, it might seem confusing that a group that dedicated itself to the relief of suffering and other food-related problems is viewed as a corporate monster. Their intentions were, as their ads claim, to improve the world through biotechnology. Their ends were the same as many socially dedicated groups, but their means were different. It is important to understand why Monsanto was singled out, and although some of it had to do with Monsanto, a lot of it had to do with the cunning of the opposition. Monsanto's big mistake, by its own admission, is that it approached the industry with too much hubris. Not knowing how to engage its stakeholders, the company thought that it could just explain that it had a superior product and the world would recognize its benefits. The company's fall is best characterized in a speech delivered by its current CEO, Hendrick A. Verfaillie:

> We were blinded by our enthusiasm. We missed the fact that this technology raises major issues for people—issues of ethics, of choice, of trust, even of democracy and globalization.
>
> We didn't understand that when it comes to a serious public concern, the more you stand to make a profit in the marketplace, the less credibility you have in the marketplace of ideas. When we tried to explain the benefits, the science and the safety, we did not understand that our tone our very approach was seen as arrogant. We were still in the "trust me" mode when the expectation was "show me."
>
> And so, instead of happily ever after, this new technology became the focal point of public conflict, the benefits we saw were jeopardized, and Monsanto became a lightning rod.[20]

---

[20]Verfaille, Hendrick A. *http://www.monsanto.com*. November 27, 2000, 3.

Although pride may have prevented Monsanto from adequately identifying its stakeholders, arrogance alone did not make the company the poster child of the anti-GM, antibiotech movement. In a way, some might say unfairly, they were singled out because they represented the best case for the opposition. Because they have become synonymous with the "frankenfood" debate, it may never have occurred to anyone to ask, in retrospect, "Why Monsanto?" They have been surrounded by the tacit assumption that they are the only arbiters of genetically modified evil, as critics might say, or of good, as proponents, though unlikely to speak loudly, might say. That dubious distinction has been levied because they are one of the few companies in this area who have consolidated their business to depend strongly on this technology. In 1995, upon his ascendance to CEO of Monsanto, Robert Shapiro began a three-year plan to spin off its chemical business and dedicate itself to biotechnology. With an emphasis on the link between agriculture and biotechnology, the company was poised to herald the new generation of life sciences, bringing the natural and the technological worlds closer than ever before; however, the company was also betting heavily on the success of this sector. This dependence did not go unnoticed by the opposition, and it made Monsanto their perfect target. Organized groups like Greenpeace make their voices heard through carefully chosen, efficient actions. By focusing on those areas where the most difference can be made alongside the loudest splash, they accomplish the dual goals of furthering their cause and drawing attention to it. Monsanto was the perfect mark, as Lord Peter Melchett, former British Labour Minister and a leader of the Greenpeace movement against biotechnology, articulates: "Of all the companies in this business, . . . Monsanto is the most committed to biotechnology. They are no worse than DuPont. But DuPont can survive without genetically modified organisms, and I don't think that Monsanto can. So we had an opportunity with them that we did not have with anyone else."[21]

Although it may seem Monsanto had all cards stacked against them, they made the mistake of exacerbating matters, as Robert Shapiro stated prior to stepping down as CEO following the 1999 merger with Pharmacia & Upjohn, "Because we thought it was our job to persuade, too often we forgot to

---

[21]Specter, M. The Pharmageddon Riddle, *The New Yorker,* 2000, p. 63.

listen."[22] In time, Monsanto would change its attitude and its policies to address social demands and issues. One of their earliest maneuvers was to avoid commercializing the Terminator technology that had become a corporate godsend because of its ability to copy-protect seeds; however, that very characteristic became fodder for the debate, as it deliberately used science to limit the agricultural freedom of those farmers, domestically and abroad, who could not afford to buy seeds annually. Interestingly, the Terminator never even made it into Monsanto's laboratories. After Gordon Conway, "head of the Rockefellar Foundation, publicly urged Monsanto to abandon the gene, . . . and even though the technology does not yet exist, and nobody can say for sure whether it would even work, . . . Shapiro announced that Monsanto would not pursue, develop or ever use the Terminator. It was among the first times in the history of science that such a prominent discovery was disavowed years before it was even clear what its value might be."[23] Conway was responding to the social baggage that this gene seemed to carry, and this move marks a significant step for Monsanto, which would begin to rededicate itself to avoiding the mistakes it had made in the past.

The spotlight on this technology has made it the subject of more than social action; legal action has also been taken against the company. As a representative of the GM industry, Monsanto was going to have to become comfortable with the notion that lawsuits, founded and unfounded, would be targeted at the company. Longtime antibiotech activist Jeremy Rifkin could not be expected to remain tranquil on this issue, and he used legal action to break his silence. Aligning himself with the National Family Farm Coalition, the class-action lawsuit revealed another way in which the industry could be damaged. According to a press release by Bill Christison, President of the Coalition, announcing the lawsuit:

> Monsanto and other life science companies have rushed GMO products to market without proper testing to determine long-term effects of these new creations. And indeed, GMO products are new creations. They are not simply the result of an evolutionary process, nor of hy-

---

[22]Ibid., p. 58.
[23]Ibid., p. 58.

bridization. Doctors, scientists, and others have raised red flags concerning the effects of this new technology.

The efforts of Monsanto and their co-conspirators has been to flood the world with seeds that produce products consumers do not want to eat. In addition, these GMOs are contaminating the entire food chain. We demand USDA to stop funding terminator type technologies and instead urge them to allocate these funds to land grant universities to better fulfill their mandate to provide public variety seeds at no cost to the farming community. Contrary to the U.S. Department of Agriculture's position that genetic engineering is necessary to feed the world; today we have a world that is awash in surplus grain resulting in record low farm prices.[24]

Although Monsanto maintains that the suit is "without merit and [the company is] confident that the suit will eventually be dismissed,"[25] there is still a cost involved for legal representation and any ensuing damage to the company's public image. The lesson is that opposition may manifest in more than just protests; those concerned with the issues are well versed in their rights, and they will exercise them to protect their interests. Indeed, much of the hostility arises out of the sense that rights are or will be violated.

After its merger with Pharmacia & Upjohn, Monsanto released a policy statement explaining its position on its technology and how the technology fit in society. No longer would they dictate to the world what it needed; rather, Monsanto would proceed based on dialogue and communication with its stakeholders. "We have reached a crossroads in the debate about biotechnology. Our experience in developed and developing countries alike has demonstrated to us and many others that this technology is safe and valuable, and that it offers benefits too important to ignore. Our experience has also taught us that people have legitimate concerns about this technology, and it's our responsibility to resolve those concerns."[26] The new policy,

---

[24] *http://www.psrast.org/usfarmlawst.htm.*
[25] Monsanto press release, Background Statement On Rifkin Lawsuit. Tuesday, December 14, 1999. *http://www.monsanto.co.uk/news/99/december99/141299_monsantobis.html.*
[26] Verfaillie, H. *http://www.monsanto.com.* 2000.

known as the Monsanto Pledge, replaced the 1990 version, which dedicated the company to sustainable development and environmental consciousness. Under the new pledge, Monsanto has five points that guide its operations and foster a new sensitivity to public concerns. These conditions are: dialogue, transparency, respect, sharing, and delivering benefits. Summarized briefly, the pledge is as follows[27]:

Dialogue
- We commit to an ongoing dialogue with all interested parties to understand the issues and concerns related to this technology.
  - To this end, we commit to create an external Biotechnology Advisory Council from a range of constituencies with an interest in biotechnology to meet, discuss, advise, and help us make decisions.
  - And we commit to involving our customers to help us make decisions about the development, use, and stewardship of new agricultural technologies.

Transparency
- We commit to transparency by making published scientific data and data summaries on product safety and benefits publicly available and accessible, and we commit to working within the rigorous, science-based regulation as required by appropriate government agencies around the world.
  - We will make both Monsanto research and external research by universities and other institutions available through the Internet and other public venues.
  - We commit our support for a mandatory premarket notification process for Food and Drug Administration (FDA) review of all biotechnology products in the United States.
  - We commit to work toward the establishment of global standards for the quality of seed, grain, and food products.

---

[27] *http://www.monsanto.com/monsanto/about_us/monsanto_pledge/default.htm.*

Respect
- We commit to respecting the religious, cultural, and ethical concerns of people throughout the world by:
  - Commercializing commodity grain products only after they have been approved for consumption by both humans and animals;
  - Not using genes taken from animal or human sources in our agricultural products intended for food or feed;
  - Never commercializing a product in which a known allergen has been introduced;
  - Using alternatives to antibiotic resistance genes to select for new traits as soon as the technology allows us to do so efficiently and effectively in a manner that has been proven safe; and,
  - Underscoring our commitment not to pursue technologies that result in sterile seeds.

Sharing
- We commit to bring the knowledge and advantages of all forms of agriculture to resource-poor farmers in the developing world to help improve food security and protect the environment.
  - To this end, we have created a dedicated team within Monsanto to facilitate technology sharing and agricultural development collaborations with public institutions, nonprofit groups, and local industry around the world.

Benefits
- We commit to work for and deliver benefits for farmers commercially as well as environmentally.
  - Environmentally, we commit to develop technology that directly contributes to a vision of abundant food and a healthy environment by:
    - Using biotechnology to promote integrated pest management (IPM) and reduce agricultural inputs, such as we have seen with the reduction of pesticides in the United States;
    - Working with growers worldwide to double the use of tillage practices that conserve soil and moisture over the next five years; and

- Ensuring that all of our products and practices protect wildlife and beneficial species.
- Commercially, we intend to launch new genetically improved commodity crops in the United States only after they have received full approval for food use and animal feed in the United States and Japan. We hope also to extend this intention to Europe as soon as it has established a working regulatory system.
- We are able to state this intention as long as there are science-based regulatory systems that make timely decisions. If the regulatory systems are not functional, we cannot allow the breakdown to deny U.S. farmers the choice of new technologies

This policy tries to systematically address the major issues that activists bring against the industry. Although meeting the goals of these tenets is a work in progress, and will continue to be so, the intentions fall short at only two critical points: the total removal of biotechnology from the company's operations and labeling its products. No one could ever expect Monsanto, or DuPont, Pioneer Hi-bred, or Novartis, for that matter, to relinquish all of its efforts in this field. Too much time and money has been invested, and, as these corporations are apt to note, too many benefits will be delivered. Furthermore, they address the issues of risk, which, scientifically, is the most that a biotechnology company can do. The arguments against biotechnology and the evidence used to support the arguments are different matters. Being against biotech in principle is irrational; however, being against the damage that biotech might levy is perfectly understandable. The ideological opposition to biotechnology can never be overturned. That is the nature of ideologies—they supply their own foundation and welcome no other points of view. They believe they are right to begin with, thus counterarguments have no place in their system of thought.

Although omitting any statements that address labeling makes sense economically, it highlights how the market restricts corporate operations. Believing that labeling food as "genetically modified" will put off consumers, Monsanto refuses to give its competitors, especially those that use genetic modification, an advantage in the marketplace. Although this attitude is understandable, it indicates two important points. First, Monsanto's commitment to its consumers does not extend as far as total transparency; otherwise, it would let its customers know exactly what they were eating. Granted, there are no standards set to define what constitutes genetically modified food products, which will be discussed in the last section, but that these products are modified at all offers enough rationale to label them in the eyes of many. Second, because competition impedes this level of transparency, effort needs to be made by all of the companies in this area to label their products and set their own standards for this practice. If one group does not label, then it may give the false impression to consumers that their products are devoid of genetic modification. A competitive advantage may be conferred based on false pretenses, and this is exactly the type of conduct that may lead to governmental regulations requiring labels. If the GM food industry intends to avoid regulatory interventions and master its own fate, then all of the players will need to convene and articulate their own standards.

Although many of Monsanto's pledges are self-explanatory, and quite ambitious, there are a few points that deserve some attention. The company sets a limit as to the standard of safety they will meet. Rather than set its own norm, which will be subject to national regulations wherever they sell products anyway, the firm states that it will release its seeds only when regulatory agencies give their approval. Regulatory agencies in the United States do not have the poor reputation that Britain's Mad Cow disease has attached to MAFF, thus the Food and Drug Administration (FDA), Environmental Protection Agency (EPA), and the United States Department of Agriculture (USDA) represent a triumvirate of trusted governmental departments. When the company states that nonfunctioning regulatory agencies in other countries do not address the issue, Monsanto will have to rely on the analysis that the United States regulators have issued. Although it may seem unclear as to whether Monsanto would release products deemed unsafe in the United States to other countries, the maxims articulated under the Respect portion of the pledge addresses this point, where the statement is made that they will commercialize "commodity grain products only after they have been approved for consumption by both humans and animals."

Another important point is the assurance that animal or human genes will not be used in the production of any agricultural products. This line of reasoning is meant to address the religious implications of the technology. Many religions forbid the consumption of certain types of animals, and it is still unclear as to whether a gene from one of these animals evokes the same proscription. For instance, does a gene from a pig inserted into corn violate kosher food practices in the Jewish faith? To obviate this problem, Monsanto will not use any animal traits in their plants.

Finally, special attention should also be paid to the Dialogue and Sharing promises. Monsanto has identified that it has to engage its stakeholders; the company has also realized that stakeholders extend beyond the farmers who buy GM seeds. By hosting a dialogue with as many interested parties as possible, Monsanto hopes to bring public concerns into its corporate operations. To to this, the company has formed a Grower Advisory Council, a collection of independent agricultural specialists who will advise on corporate strategy. The group's action items are covered in the Sharing portion of the pledge, which Monsanto fulfills by sponsoring research, both internally and through independent funding to unlinked labs, to determine the scientific risks involved with the research. In another Dialogue effort, the company has initiated a movement dedicated to educating the public on biotechnology. Through teaching grants and a website dedicated to biotechnology, known as the Biotechnology Knowledge Center (*http://www.biotechknowledge.com*), Monsanto intends to present the details surrounding the science and the ethics of GM foods. The website is already aiding Monsanto by hosting objective reports such as "Genetically Modified Crops: The Ethical and Social Issues." Created by the prestigious and respected bioethics center, The Nuffield Council on Bioethics, which is chaired by moral philosopher, Alan Ryan, this report, whose objectivity is not in question given its fair and distinguished source, states that "*The Working Party does not believe that there is enough evidence of actual or potential harm to justify a moratorium on either GM crop research, field trials or limited release into the environment at this stage. . . . The Working Party concludes that all the GM food so far on the market in this country [Britain] is safe for human consumption.* [italics theirs]"[28]

---

[28]The Working Party, Nuffield Council on Bioethics. Genetically Modified Crops: The Ethical and Social Issues. May 27, 1999, p. 4.

Although it remains to be seen if these efforts will help Monsanto's reputation, the question that many of the company's competitors want answered is whether the efforts will help the industry. The backlash against GM foods and biotechnology has not helped the causes of those firms that occupy this space. In an article in the *New Scientist,* this position was explored. "Monsanto, the American biotech giant," the article claimed, "is facing an unprecedented wave of criticism from within the industry. Many of Monsanto's rivals say the company is largely to blame for a consumer backlash that could cripple the prospects for genetically engineered food."[29] It further reports on the specific attitudes of some of the larger biotechnology companies producing genetically modified seeds. "'We have a PR mountain to climb,' says Willy de Greef, head of regulatory and government affairs at Novartis Seeds in Basel, Switzerland. 'You have a problem if the market leader has firmly set ideas about how to do things, which others might not agree with,' he adds. 'An expensive failure can be made into an asset if you've learnt from it, but Monsanto still has some learning to do.' Zeneca, the British-based biotechnology giant, also feels aggrieved, not least because it won applause from consumer groups in 1996 by labeling its tomato puree as containing genetically modified tomatoes. 'It's a matter of respect for your customer,' says Nigel Poole, head of regulatory affairs at Zeneca Plant Science in Bracknell, Berkshire."[30] Of the many lessons to be learned from Monsanto's approach, this is one that should not be forgotten: negative public perception can extend beyond a single company to any and all firms that develop the scrutinized technology. Even privately held firms are suffering as they try to raise funding for their ventures. According to *The Economist,* "The public furore over genetically modified (GM) foods—particularly fierce in Europe—has not helped matters. Berry Summerour, an analyst with Stephens Inc., an investment bank, finds that some British fund managers simply do not want to hear the words 'agricultural biotechnology.' Those who do will sometimes not admit to it: for example, Paradigm Genetics, a genomics firm, has two large British investors, one of which prefers to keep its interest quiet."[31]

[29]Mutiny against Monsanto. *New Scientist,* October 31, 1998, *http://www.newscientist.com/gm/gm.jsp?id=21580200.*
[30]Ibid.
[31]Dry Season. *The Economist,* November 2, 2000.

The consequences of ignoring the public resistance to technology are not limited to image and lawsuits. When Monsanto's products were accidentally shipped alongside standard soya, the result was not as economically debilitating as a recent event in the United States. Aventis CropScience endured a public relations nightmare when StarLink Corn, a grain modified to exhibit pesticide qualities, had been detected in human foods. The corn was designated for animal feed, but inadequate segregation methods had resulted in StarLink use in commercially available human products. Suspicious of potential mix-ups, environmentalist group Friends of the Earth sent a series of products sold by Kraft Foods to Genetic-ID, an independent DNA testing company. Through a series of tests, Genetic-ID discovered the presence of the *Bt* gene in the products, indicating that GM food must have been used in the production of Kraft's taco shells and other products. There is an irony here in that human and animal food may have been mixed in the past, and no one was any the wiser if it had; however, the introduction of this controversial technology provided both the basis of segregation and the means of detection. Ultimately, Kraft had to recall a number of its products, while Aventis did the same with its crops. All tallied, the cost to the biotech firm Aventis are estimated to be at least $500 million, and Kraft is still adding its numbers.[32] Had the company responded to the ethical concerns that surround the industry, this mishap may not have damaged the company's bank account or its reputation; however, Aventis did not make the effort of demanding segregation between its GM crops and standard crops. Although the company probably expected that farmers would designate different silos for each type of grain, Aventis still did not remain invested in its products long enough to ensure adequate distribution. Although the *Bt* toxin is not hazardous to human health (the EPA has approved *Bt* pesticides for over 20 years), there have been reports of allergic reactions to these transgenic products. It is unlikely that there is enough foreign genetic material in processed foods (processing removes many genetic and proteomic traits from food) to elicit an allergic response; however, the FDA has received complaints of individuals having allergic reactions to the tainted products.[33] Not ending

---

[32]Biochips Down on the Farm. *The Economist,* March 22, 2001.
[33]Goodman, Troy. Should You Fear Franken-Corn? CNN.com. November 10, 2000, *http://www.cnn.com/2000/FOOD/news/11/10/starlink/.*

with these health reports, the issue took another form when a series of United States farmers filed a class action lawsuit against the biotech giant. Alleging that Aventis "failed to inform farmers that StarLink had been approved by the Environmental Protection Agency for use only in animal feed and for industrial purposes out of concern that a protein in the genetically altered corn might set off allergies in humans," the farmers contend that "the crops of many corn growers who did not plant StarLink were contaminated by StarLink corn and the subsequent crisis in the nation's grain-handling system closed off foreign markets and depressed the price of American corn here and abroad."[34]

As the farmers' negligence suit indicates, the repercussions of mistakes in the handling of GM foods reach far. A variety of people are reacting, not to mention entire nations. International mergers and acquisitions are at an all-time high,[35] which means that the restrictions of the most limiting entity in the merger and acquisitions exchange will have to become standard operating procedures for the whole of the parent company when it is developing controlled technologies. Furthermore, international trade of transgenic produce has been dealt a profound blow, given that many countries will not allow the introduction of GM products into their food supplies. The clear implication to trade is reflected in the impediment of international regulatory harmonization; when dealing with foreign countries and companies, regulatory restrictions become the bottleneck for commerce pathways. Many countries already have specific ethical regulations that govern their operations. For instance, the exportation of StarLink corn to Japan, a country that depends on importation for its corn supply and does not permit the importation or distribution of GM products, was approved because the U.S. tests for modifications came back negative. "The discovery of StarLink in corn samples has raised concerns that the test plan would not be sufficient to comply with strict Japanese legislation to be implemented in April against unapproved genetically modified products."[35] Since the United States is responsible for 72% of

---

[34]Barboza, David. Negligence Suit Filed over Altered Corn, *The New York Times,* December 4, 2000.

[35]PriceWaterhouseCoopers. Pharmaceutical Sector Insights: Analysis and Opinions on Merger and Acquisition Activities, Interim Report. 2000

[36]Japan Corn-buyers Play Down New StarLink Discovery. Reuters, 02.02.01. *http://www. forbes.com/newswire/2001/02/02/rtr176721.html*

the international corn trade, and since Japan is one of the United States' largest importers of corn,[37] failing to meet these and other international requirements can severely damage the trade relations on which the GM food industry depends for its modified *and* unmodified products. Adding to the concern is the economic cost that "trusting" the exporter might bring. As the example with Japan shows, additional testing may be necessary to confirm the presence or absence of an inserted gene. If it is necessary to retest samples upon their introduction into a new country, an expense that should be unnecessary, then who will bear the costs? The importer who needs confirmation, or the exporter who has a history of reporting false negatives? Although many companies may disagree with the claims that environmentalists and other activists make regarding the scope of these issues, given these revelations, no one can continue this debate without admitting that, regardless of his or her ideological position, the implications surrounding GM foods resonate at a global level.

VI. Recommendations

Because of the widespread reputation of GM foods, there are no magic bullets that will immediately aid a company to be completely bioethically aware and identified by the public as such. The concerns about the technology are profound, and the public is not likely to trust large, multinational corporations when the companies claim that their products are better solutions to health, environmental, and social problems simply because these goods are manufactured with the aid of biotechnology. In many ways, this attitude is the problem, but it is reasonable to ask the skeptical question, How can these companies change their attitude when their value is derived solely through biotechnology? The answer is as simple and difficult as "listen"—simple because the voices of protest cry out so loudly that they are hard to ignore, but difficult because so many of the demands are unreasonable or lack a fundamental understanding of the science behind this technology.

---

[37]Ono, Yumiko and Kilman, Scott. Japan Asks that Imports of Corn be StarLink Free. *The Wall Street Journal.* October 30, 2000. A6.

Although there are a number of tools that will be introduced in this section to guide a GM food company toward being more aware of bioethical concerns, there are some general considerations that ought to be addressed. It is somewhat late in the game for "I told you so" and "what they should have done," but these insights are still relevant. Many experts have stated that the major problem behind GM foods was their choice of first-generation products. Corporations decided that certain products would be developed, but their benefits were negligible by the standards of most end users—environmental advantages and easing the workload of farmers are not tangible at the supermarket. Accoring to Robert Lipton, an economist at the University of Sussex's Poverty Research Unit, "If Monsanto had spent a lot of money and produced an egg with no cholesterol, I just don't think we would be having these problems today. . . . I always say that electricity is a fantastic invention . . . but if the first two products had been the electric chair and cattle prod, I doubt that most consumers would have seen the point."[38] The best way to naturalize GM foods into mainstream consumer products is to develop foods that meet a consumer "pull," rather than relying on a corporate "push."[39]

Another consideration is the tension between the product and the process. Most biotech firms will defend GM food by saying that genetic modifications have been applied for decades, and if seed selection is included, millennia. However, the opposition rallies around the potential, negative effects that these more efficient methods might unexpectedly deliver. More precisely, the public's sense is that corporate intentions, something that has always been suspect to activists, are wholly profit driven; thus, multinationals are only committed to aiding societies only to the extent that is most profitable. While it is somewhat naive to believe that corporations will begin operating at a loss to prove their allegiance to social ideals, it is not unreasonable to reformulate those ideals. For too long, companies have touted the advantages of biotechnology, which is an attitude that has to change. In the case of GM foods, nutrition, environment, hunger, and health have been the focus of the products, but the perception has been that biotechnology has

---

[38]Specter, M. The Pharmageddon Riddle, *The New Yorker,* 2000, p. 61.
[39]McHughen, Alan. *Pandora's Picnic Basket.* New York: Oxford University Press, 2000.

been the primary issue, whereas these benefits are incidental. The simplest remedy—simple to state, not to implement—is to focus on these issues and make biotechnology a means to these ends, rather than vice versa. Many times the criticism has been made that there is enough food in the world today to feed the hungry, yet multinationals are not helping the excess provisions reach the malnourished. Of course, it is not so simple as just sending food via the post office (which many activists would naively have the public believe), but perhaps efforts should be made to help build the infrastructure that would allow such transactions. Granted, there are many logistical difficulties, ranging from trade barriers to storage facilities, but participating with those groups that hope to meet these goals, be it monetarily or bureaucratically, would convey a dedication to these causes. Already, many pharmaceutical companies are making drugs available to developing countries, and many biotech companies are providing GM seeds, which are all steps in the right direction. However, there are more categories than just world hunger that opponents to biotech have used as evidence to substantiate the claim that firms are not only unconcerned with ethical issues, but against ethical conduct.

One way to begin the attitude change is to implement a policy framework, similar to Monsanto's Pledge, which explicitly states a company's commitment to social and environmental issues. Although many points should be considered in developing the policy framework, most of the decisions should be based on the outcome of discussions over dialogue and transparency. The exercise in formulating the policy should be inclusive of management (depending on the size of the company), and one of the earliest goals should be deciding how open to the public the firm should be about its operations. After a sketch of the policy has been made, it should be tested against the community. Because the community is quite pluralistic in its views, the company would have to carefully consider how comfortable it is with inviting outside criticism from certain factions. Perhaps the American Farm Bureau Federation would be constructive in the corporation's view, whereas Greenpeace might not. Of course, to reach maximal transparency, the most oppositional groups would be the first to be invited.

A useful tool, and a dynamic one in that it should not be abandoned after a policy has been implemented, is the PEST methodology, also known as

STEP when a more positive sounding anagram is preferred. To help frame a policy, Political, Environmental, Sociocultural, and Technological factors should be weighed. According to this analysis, there are a number of forces affecting a business, to which some respond and some do not. Thinking in terms of these characteristics will aid in understanding the specific pressures that bear upon a corporation. In their examination of business processes and the environment, Alasdair Blair and David Hitchcock identify five sources of pressure that can be viewed through this filter: direct business causes, environmental change relating from business activities, natural environment, social change related to business activity, and social change unrelated to business activity.[40] By using this framework as a heuristic tool, the range of issues that have or will emerge in reference to a technology can be identified and approached through corporate strategy.

Before presenting more tools, it is important to note that their utility depends on a thorough understanding of whom they are serving. On the one hand, there is the corporation, and on the other there is everyone else who is affected by the corporation. As daunting as the size of the latter category might seem, it is exactly its underestimation that has led to all of the problems associated with GM foods. Put simply, a technology's range of stakeholders is much broader than ever before. With information technology reaching new heights, there is very little that cannot be discovered about a company and its technology. The disappointment is, however, that much of that effort is devoted to finding evidence to undermine technological progress, rather than educate oneself on the benefits. Because of this openness of information, or forced transparency, companies have no choice but to identify their stakeholders and their corresponding interests when developing products. Even if the result is disagreement over particular stances, preparing rational justifications in advance of the criticism places the company in a less defensive, more sincere position. Again, Monsanto serves as the example; the company responded to criticism with an arrogant tone rather than a dialogic invitation. When Monsanto finally identified this mistake, former CEO Robert Shapiro recanted: "Our confidence in this tech-

---

[40]Blair, Alasdair and Hitchcock, David. *Environment and Business.* New York: Routledge, 2001, p. 100.

nology and our enthusiasm for it has, I think, been widely seen—and under-standably so—as condescension or indeed arrogance."[41]

One of the first steps, and an ongoing concern, in technology develop-ment should be stakeholder identification. In ethical philosophy, this comes through in contractarian ethics. According to this theory, moral foundations are built upon the mutual consideration of individuals who agree on the types of constraints and freedoms that ought to be realized. All of those who participate form a contract to avoid harming each other, while doing their best to help each other; more precisely, all contracted agents agree to act morally. In her essay "Duties Concerning Islands," moral philosopher Mary Midgley discusses the importance of understanding how inclusive or exclu-sive individuals should be regarding the entities with whom they enter into these contracts.[42] Her analysis offers a good starting point for companies in that she lists a range of agents who should be considered as participants in the pact. Although not all of them will require consideration, the categories might help companies decide how to identify stakeholders.[43] See Table 2-1, in which the categories are listed in no particular order.

In this analysis, Midgley does not contend that this list is comprehensive, nor does she believe that all of these stakeholders require consideration in every moral decision: "The point is this; if we look only at a few of these groupings, and without giving full attention, it is easy to think that we can include one or two as honorary contracting members by a slight stretch of our conceptual scheme, and find arguments for excluding the others from serious concern entirely."[44] The lesson for biotechnology is to begin with the broadest range of stakeholders and address the interests of those who have not been rationally excluded from consideration. Those that remain form the foundation of the ethical contract.

Safety is also an issue that requires strict attention. In the United States, there are three regulatory agencies responsible for the distribution of GM

---

[41]How the Mighty Fall. *The Guardian.* November 22, 1999. *http://www.guardianunlimited. co.uk.*

[42]Midgley, Mary. Duties Concerning Islands. In *Ethics,* Peter Singer (Ed.), New York: Oxford University Press, 1994, pp. 374–387.

[43]Ibid., pp. 381–382.

[44]Ibid., p. 382.

Table 2-1

Categories	
Human Sector	1. The dead 2. Posterity 3. Children 4. The senile 5. The temporarily insane 6. The permanently insane 7. Defectives, ranging down to "human vegetables" 8. Embryos, human and otherwise
Animal Sector	9. Sentient animals 10. Non-sentient animals
Inanimate Sector	11. Plants of all kinds 12. Artifacts, including works of art 13. Inanimate, but structured objects - crystals,
Comprehensive	14. Unchosen groups of all kinds, including families and species 15. Ecosystems, landscapes, villages, warrens, cities, etc. 16. Countries 17. The bioshpere
Miscellaneous	18. Oneself 19. God

foods: the USDA, EPA, and FDA. Under the Federal Plant Pest Act, the USDA is charged with overseeing the experiments, such as field trials, that ensure the safe release of plants and seeds, as well as making certain that the crops do not damage the environment. The EPA is given jurisdiction by the Federal Fungicide, Insecticide and Rodenticide Act (FIFRA) and the Federal Food, Drug and Cosmetic Act (FFDCA) to evaluate the pesticidal and herbicidal properties of the modified plants. Finally, the FDA, also by the FFDCA, assesses the safety of all human and animal food. Most countries have their own regulatory system, which are viewed with varying degrees of credibility. Thus, corporate biotechnology may wish to perform their own experiments and risk analysis. In general, sponsoring independent research, as well as performing in-house trials, to address the particular concerns that have been raised would be an excellent starting point. As the issue is not about the process (biotechnology), but the effects, the focus should not linger over just GM food. There are many different plant breeding techniques, and all of them are subject to similar regulations. In some cases, the crops do not have to go through as many regulatory agencies as GM plants, but the overlap justifies an effort on behalf of seed companies, *not just GM companies,* to support more research to gather data on allergenicity, ecological effects of pesticides, as well as other areas of risk in crop biotechnology.

Beyond general experimentation, risk analysis must be an integral part of any company's technology development if the firm is producing genetically modified plants. Risk, it should be noted, is a term that characterizes the possibility of some type of an unwanted event. The precautionary principle is often invoked in these discussions as the most commonsense approach to risk analysis. According to this principle, a product should not be released until any and all of the negative effects are found to be absent, and in the case of uncertainty, a company should err strongly on the side of precaution and not release the product. Thus, no product will move beyond a controlled setting until it has met a defined measure of safety. Every contingency within a reasonable limit must be examined, and it is critical to understand that the product must meet some threshold for the experiment to be valid. For instance, perhaps the allergenic effects of an inserted pesticide gene is under scrutiny; if the product, say a tomato, does not elicit an allergic reaction, but the tomato does not produce the pesticide, then the experiment does not meet the standard of success because it failed to perform as expected. Due to the importance of these types of analyses, the rigor of the corresponding experiments must be uncompromising if the company chooses to commit to the principle.

Although the precautionary principle may be too extreme for some tastes, there are general elements that must be observed in risk analysis, which then should extend to risk management during all trials, regardless of the chosen method. "The utility of risk analysis derives from developing a rational framework whereby the knowledge-based description of risk (a science driven process) is integrated with social, cultural, economic, and political considerations to manage and communicate risk in policy decisions and implementation."[45] Much of this discussion may seem familiar, since risk analysis is nothing new to biotechnology, particularly agbiotech. Rather than develop new techniques to perform risk assessments, the focus should be on the strict maintenance of such procedures and communicating these findings to

---

[45]Wolt, Jeffrey D. and Peterson, Robert K. D. Agricultural Biotechnology and Societal Decision-Making: The Role of Risk Analysis. *Dow AgroSciences,* January 1, 2001, *http://www. biotechknowledge.com,* p. 3.

the public. "The precepts of risk analysis are by no means unknown within agricultural biotechnology. These precepts, however, have not been clearly evident as agricultural biotechnology has moved into the marketplace. Attempts to deal with public sentiment against this technology initially resulted in a 'trust us, it's safe' approach to risk communication. However, the public was faced with insufficient information to understand what was meant by 'safety,' a judgement."[46]

This last quotation touches upon another major theme of bioethics and biotechnology—marketing. Most ethicists and scientists would agree that using ethics for the purposes of public relations is antithetical; however, companies have an obligation to reverse the structure and use public relations for ethics. Although there are obvious advantages to conveying ethical attitudes and behavior via PR, for instance, it might yield a competitive advantage in a setting where the number of consumers who choose products based on the social responsibility of the producer is on the rise,[47] it is necessary to help set an example for the remainder of the industry. According to Richard Welford, Professor of Corporate Environmental Management at the University of Huddersfield and Professor of Sustainable Management at the Norwegian School of Management, marketing is integral to such a process. Concerned with sustainability, a strong analog of bioethics in this case, Welford states, "what is very important for a company that wants to move in the direction of sustainable development is the ability to differentiate its products and its overall corporate image. Therefore, the marketing function is vital if the organization is to communicate its difference. However, there is a clear win–win situation here if, at the same time communicating the difference, the company can also persuade and educate its consumers and the wider public to act in more environmentally friendly and socially responsible ways."[48] To do so, however, requires more than commitment, it requires a system that provides the information that can direct these maneuvers.

To fully realize the bioethical vision within a marketing system, the task

---

[46]Ibid., p. 4.
[47]Dalla Costa, John. *The Ethical Imperative.* Reading, MA: Perseus Books, 1998.
[48]Welford, Richard. *Corporate Environmental Management, 3.* London: Earthscan, 2000, p. 109.

requires a new emphasis on data acquisition.[49] Since there are a multitude of categories that demand attention, each has its place within the decision-making process. By broadening the flow of information, such interests can be addressed. Data should be gathered from three sources: primary, secondary, and through alliances. Relying on direct research, primary sources should engage stakeholders through surveys and other "face-to-face" procedures to determine attitudes and market trends. Secondary sources, such as trade publications, must also be included. Both journals *Science* and *Nature*, for instance, dedicate a number of pages to social issues/commentary each week, which are timely, topical, and informative. Finally, cooperative alliances with organizations, both supporters and antagonists, offer another form of information. These groups may perform their own analyses, not only providing useful, cost-effective information, but also giving a different perspective than the one to which a firm may be accustomed.

With this data, the next step in the information system can be taken. These facts and figures ought to be disseminated into a workable format; for instance, placing them in the following categories: stakeholder information, impact information, and technical information. Stakeholder information has been covered and technical information seems self-explanatory (risk data, current science, patents, etc.), but impact information deals with the social and technical implications of the research that does not fit into the other two categories, as well environmental damages/benefits that may result. Critical at this juncture is to analyze these findings and find a way to implement the results into corporate operations. Furthermore, the results should be updated regularly, and kept in an accessible, centralized area or library. Although the temptation may be to jump the gun and go right to marketing, that would not serve the interest of the system, which is to serve bioethics. Since public relations and advertising campaigns ought to reflect the company, this intermediary step must be taken. Furthermore, the information gathered should not only function as an educational measure for the company; it should provide the means of communicating the information to the public through marketing strategies. It is clear that selling does not

---

[49]This scheme and methodology, visually depicted in Figure 4.1, is adapted from, Welford, R. *Corporate Environmental Management, 3,* 2000.

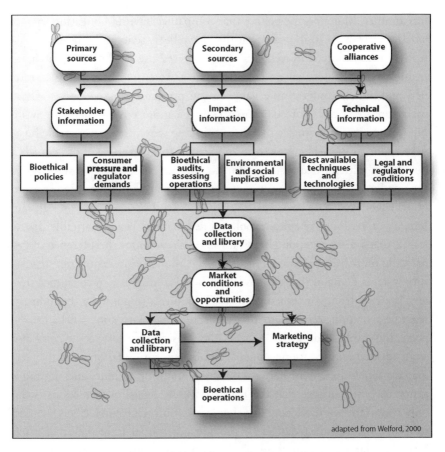

Figure 2-5   Bioethical information management system.

work in this industry; many companies have tried and failed to persuade the public. First educate, then sell. This affords the public the most valuable commodity that the industry can offer: informed choice. For a visual representation of the bioethical information system outlined in this section, see Figure 2-5.[50]

Although educating the public through marketing initiatives is one mode of assuring informed choice, the hotly debated topic of labeling products of-

---

[50]Welford, R. *Corporate Environmental Management, 3,* 2000.

fers another. Many people believe that companies have an obligation to post labels on products to indicate that they have been genetically modified. Already, the antibiotech movement has introduced the notion of "negative labeling," which states that no GM techniques were used in growing a product, and which confers more value to the product in some cases. "These social pressures have forced businesses to be more environmentally aware but at the same time to see that such thinking may offer considerable business opportunities. The growth in the market for organic food offers such a possibility, with a price premium for the producer but more particularly for the vendor."[51] At the same time, consumers are applying pressure for more knowledge; according to a recent article in *The Wall Street Journal*, "In late January, a national telephone poll funded by the Pew Charitable Trusts found that 75% of respondents wanted to know about the presence of genetically modified ingredients in food."[52] Although some countries' regulatory systems demand that the foods be labeled in response to consumer anxiety, pressure from concerned groups will require a response by the industry regardless of policies, foreign or domestic. Despite labeling's negative aspects, the public is unlikely to accept the considerable cost to the company as an adequate justification against it. Public perception falls toward the feeling that labels are absent because there is something to hide, and the more generous critics believe that the negative public image of GM foods is the only reason for not disclosing this characterization. Currently, labels are used to either warn against danger or provide nutritional information. United States regulations require food labels when new products are not "substantially equivalent" to their counterparts. When the products pass through the EPA, USDA, and FDA, the measure of substantial equivalency is imposed, and should the product fail, it will require a label indicating the difference. Placing labels on products when it is unnecessary with respect to the regulations may raise a red flag because it might confuse the process; however, public demand cannot be ignored. When the opposition cries loud and rationally enough, the government will respond, and the opportunity now is

[51]Blair, Alasdair and Hitchcock, David. *Environment and Business*. New York, Routledge, 2001, p. 125.
[52]Callahan, Patricia and Kilman, Scott. Seeds of Doubt. *The Wall Street Journal*, April 5, 2001, A6.

to anticipate legislation prior to enacting it, rather than catching up later (see Figure 2-6).

There are different types of information to consider including in a label. Obviously, some indicator like "improved through biotechnology" has to be included to point out the use of GM technology. The most important aspect, and something on which the FDA will insist, is honesty and clarity—nothing should be stated that is false or confusing. Also, a scientific threshold can be established if corporations believe that the food processing method has removed any of the inserted DNA and its traits. Some measuring system must be implemented, and the method should be communicated via the label, either directly or through a reference to another source; this is an opportunity to educate that should not be squandered—either a brief description of the technology or encouragement to visit a website or request information via mail should be included. The process can be used to benefit the company, particularly when products are introduced that have direct benefit to the consumer.

Finally, an advisory board equivalent to an Institutional Review Board (IRB) should be implemented. An IRB consists of a panel of experts regarding science, ethics, and other areas who advise researchers on responsible

"Sorry, 'Ms. Know-it-all', but there's nothing listed here about genetically modified foods."

ONORATO

Figure 2-6.

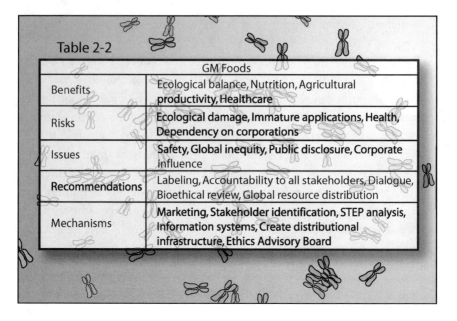

Table 2-2

GM Foods	
Benefits	Ecological balance, Nutrition, Agricultural productivity, Healthcare
Risks	Ecological damage, Immature applications, Health, Dependency on corporations
Issues	Safety, Global inequity, Public disclosure, Corporate influence
Recommendations	Labeling, Accountability to all stakeholders, Dialogue, Bioethical review, Global resource distribution
Mechanisms	Marketing, Stakeholder identification, STEP analysis, Information systems, Create distributional infrastructure, Ethics Advisory Board

modes of research. The group should be given specific responsibilities that explore both ethics and corporate goals, and representation on the board should be varied. Obviously, geneticists, physicians, farmers, and bioethicists/philosophers should participate, as well as environmental scientists and other technical specialists. There is another, less obvious group that should be included as well: anthropologists. This recommendation may seem out of place, but the reason is very clear. Until now, the entire discussion has focused on the First World, where the luxury for choosing food is often taken for granted. Abroad, there are sociocultural–economic issues that differ from the familiar terrain that, comparatively speaking, we are very lucky to walk. Anthropologists are trained in the study of other cultures, and better than any other discipline, this one offers an informed look at those foreign cultures into which the technology might be introduced. To ensure that all of the freedoms and privileges afforded domestic markets are extended to the developing world, corporations will need a guide who can bridge the cultural gap to avoid an unnecessary and possibly dangerous compromise.

Genetically modified food is the harbinger of great hope and promise for the world, but, like a number of new technologies, many meet it with apprehension. By ignoring those people who criticize the technology and demand assurances of safety, the industry has dug a hole for biotechnology, not just GM technology. Unless there is a sustained and considerate response, the industry may harm itself even more. Should this pattern continue, governments, corporations, and activists alike will come down hard on the groups that do not learn from others' mistakes, and in more ways than one, GM companies will regret having to reap what they have sown.

3
DNA DATA BANKING

What is robbing a bank compared to founding a bank?
—Bertolt Brecht, *Die Dreigroschenoper*

I. Executive Summary

With the completion of the Human Genome Project (HGP), the postgenomic era begins. Although corporations and scientists deliberate on how to leverage this information, many have realized that the HGP's effort has only provided one part of the promise it offered upon its conception. To discover disease genes and other inheritable traits, bioetechnologists are amassing large storehouses of DNA. By analyzing patterns of disease inheritance in these samples, information can be accumulated and stored to reveal which genes are involved with disease.

The Science

DNA data banks collect tissue or DNA samples to extract and store information derived from them. By creating banks for specific diseases, the genetic commonalities can be identified and associated with diseases or disease states. In other scenarios, genetically homogeneous populations can be used to search for differences in otherwise similar genomes. With this strategy, the variations would stand out, facilitating the discovery of disease-linked genes.

The Benefits

Studying the genetics of disease will aid in the discovery of new therapeutics. A better understanding of the genetic basis of disease will also help treat the disease at many different levels. For instance, pharmaceuticals can be developed to overcome the symptoms of disease, while gene therapy can be used to "fix" the defective gene.

The Issues

Individuals concerned about the repercussions of DNA data banking generally agree that the danger associated with this technology centers on conceptions of privacy. Information revealed about a person's genome may lead to discrimination, and unless confidentiality is strictly observed, participants in a data bank may face substantial repercussions for donating their DNA. Information contained within banks may be used to compromise donors' insurability and employability.

The Industry

To better understand how these issues affect industry and how industry affects society, DNA Sciences is used as an example to explore the technology. Although the standard concerns over privacy affect this company as much as any other, the reliance that DNA Sciences has placed on the Internet as a recruitment tool as well as a communication medium with donors further complicates matters. The methods by which DNA Sciences deals with the privacy challenge levied by both DNA data banking and the Internet offer an excellent starting point to dissect these issues.

Recommendations

Although numerous steps can and should be taken to ensure privacy, the most emphasis is placed on the informed consent process. Because informed consent is a standard part of a researcher's procedures, it offers the perfect legacy system in which to integrate the safeguards to ensure confidentiality. Other ideas are also discussed to relate the argument to the Internet; for instance, file protection and encryption methods to better protect the donor and the company from the repercussions of privacy violations.

Of the many accomplishments made by modern science, decoding the human genome ranks high among them. Identifying genes associated with disease has opened a new era of medicine, in which genetic information will be used in every clinical stage of an illness, from diagnosis to treatment. More precisely, this knowledge will allow researchers and physicians to determine a person's probability of suffering from any genetic disease, and if that probability is high or certain, a genetic profile will also be used to prescribe the exact drug cocktail or therapeutic regimen to which the patient will best respond. With vast efforts like the Human Genome Project, scientists have already begun the groundwork that will eventually lead to the full realization of this promise.

Before humanity achieves this scientific goal, more people than scientists will have to participate in the necessary experiments. Although much of the technological know-how already exists (and will receive attention in Chapter 4), it is relatively useless without an immense number of tissue samples on which experimentation can be performed. Without a large cache of different DNA samples, scientists cannot address the questions whose answers will constitute each rung in the ladder leading toward a healthier world. Amassing these samples and the genetic information they yield is the scientific and corporate goal of DNA data banks—central locations where many thousands to millions of genetic profiles will be stored after donors "deposit" their tissue in the bank for scientific analysis. Possessing the resources that will make the analysis possible, companies are building storehouses of DNA collected from populations spanning the globe. Through these efforts, corporations will gather enough data to aid in the fight against genetic disease while profiting by these efforts. On the surface it seems like a win–win proposition. One group will make money in exchange for the work it puts in, and another group will see cures for diseases in exchange for the genetic material they donate. How could this business possibly fail? Why would anyone question the ethics?

Perhaps this is too simplistic; few business minds will ever claim that there is such a prospect as a "sure thing." What makes this scenario a little more challenging is that traditional business concerns are no longer the only agents working against success. When dealing with biotecnology, individuals become wary of how their tissue might be exploited. They have concerns that are echoed in the ethics of research that, in the past, were just esoteric points far removed from the practical world of business. Today, however, the populace has become cautious of both corporations and biotechnology. Under these circumstances, one half of the aforementioned business equation—those people who donate genetic material to corporate research—are no longer constants. They are variables in the biotechnological calculus, and they need to be assured that their interests are met prior to experimentation, as well as after. To ignore this fact is to compromise the foundation of this research. People will be unwilling to relinquish their DNA to any group that makes only faint attempts to protect the donor. To fully understand this outlook, corporations need to recognize the importance of bioethics to their daily operations.

Tissue banks, which are in effect no different than DNA banks (biological tissue is a source of DNA), are usually very strict about their policies regarding collection and use. Furthermore, because DNA and tissue are derived from human sources, they often enjoy the same legal and ethical protections as human subjects, although there are some caveats.[1] However, laws and ethical guidelines are *indicators* of acceptable conduct, not guarantors (laws explain obligations and repercussions for violating obligations, but they do not guarantee that irreparable damage will not be done). Banks must heed both if they are to proceed without repercussion, which can be substantial. In the United Kingdom, recent events have compromised the public trust in tissue donations, negatively impacting research. A pathologist at Alder Hey, a Liverpool Hospital, "had stripped and stored every organ from every child whose parents allowed permission for a post mortem."[2]

---

[1]For instance, in the United States, so long as the material remains identifiable with respect to its source, it is treated as a human subject as defined by the federal regulation known as the Common Rule, 45 C.F.R. pt. 46. If it is unlinked, that is, if the source is unidentifiable, its status is no longer equivalent to human subject research.

[2]Adam, David. Tissue Donations Slump after Revelations about Misuse. *Nature, 409,* 655, February 8, 2001.

The pathologist in question proceeded against standard ethical procedures, which require the next of kin to grant permission for the actual retention of tissue for research purposes. Although the transgression is amplified by an investigation's revelations that showed that these samples had not ever been used in research[3], the damage had already been done. Parents felt violated, and as clinicians from all spheres tried to distance themselves from the incident, the international recognition of the event affected unaffiliated research institutes. Because similar events had resulted in the decline of sample use in France and Scandanavia, "other pathologists are facing a crisis in public confidence that may lead to a shortage of donated tissue for research. . . . The political fallout from the episode has also led to the immediate suspension of some research projects, with researchers unwilling to continue until ethical and legal guidelines are clarified."[4] In this instance, a lack of ethical guidance has led to indecision on the part of both donors and researchers, ultimately impeding scientific research. It is critical then, for the sake of public, academic, and corporate interests, for research institutes to understand the ethics surrounding DNA data banking and proceed from this framework.

II. The Science

DNA data banks are closely linked to both DNA banks and tissue banks. The historical evolution of biological sample repositories explains this connection. The first institutions, tissue banks, were maintained primarily for pathology research. Established well before DNA analytic techniques (specifically, DNA extraction and sequence analysis methods), tissue banks eventually became DNA banks upon the introduction of such techniques. The leap to data banking occurred when computing tools and database management created an efficient means of storing the information derived from and linked to the DNA.

---

[3]Royal Liverpool Children's Inquiry. *http://www.rlcinquiry.org.uk.*
[4]Adam, David. *Nature, 409,* February 2000.

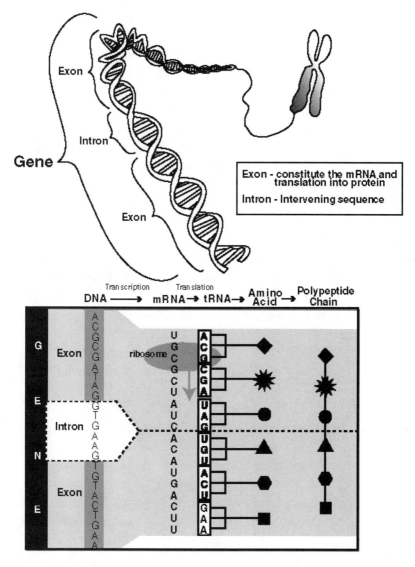

Figure 3-1   Deoxyribonucleic acid (DNA) from chromosome to base pair.
http://www.nhgri.nih.gov

As their close relationship may indicate, there is a fine distinction between tissue banks, DNA banks, and DNA data banks. As mentioned above, tissue banks house biological material either extracted during medical procedures or donated for specific research purposes. Because DNA is extracted from human tissue, there is little use in distinguishing among the two in this discussion. This approach is not at all a departure from conventional thought. According to the National Bioethics Advisory Commission (NBAC), "'human biological materials' is defined to encompass the full range of specimens, from subcellular structures such as DNA, to cells, tissues (e.g., blood, bone, muscle, connective tissue, and skin), organs (e.g., liver, bladder, heart, kidney, and placenta), gametes (i.e., sperm and ova), embryos, fetal tissues, and waste (e.g., hair, nail clippings, urine, feces, and sweat, which often contains shed skin cells)."[5]

Often stored following surgical excision, tissue is also warehoused at the request of donors. For instance, individuals with a particular disease may donate their material in hopes of aiding research on that disease. Other reasons motivate donors as well. For instance, maintaining a genetic record for identification purposes in case of accidents or tragedies. In the United States alone, these efforts have led to the accumulation of many millions of stored samples, and these samples continue to be distributed to academic and private institutions for a variety of research purposes.[6]

DNA data banks represent a slightly different phenomenon because they are not storehouses of biological material; rather, they deal in something less tangible, but potentially more insidious: information. DNA data banks are "repositories of genetic information about individuals obtained from the analysis of DNA samples . . . typically, however, they involve the routine storage of genetic information about large numbers of people; the information is generally maintained with individual identifiers and in

---

[5]National Bioethics Advisory Commission. Research *Involving Human Biological Materials: Ethical Issues and Policy Guidance,* p. 13. August, 1999. Volume I, Report and Recommendations of the National Bioethics Advisory Commission. Rockville, MD.

[6]For a detailed survey of tissue banks in the United States, see Eiseman, Elisa, Stored Tissue Samples: An Inventory of Sources in the United States. In *Research Involving Human Biological Materials: Ethical Issues and Policy Guidance.* Volume II, Commissioned Papers. Rockville, MD. National Bioethics Advisory Commission, D1–D52. January, 2000.

computerized form, making it easy to access and (potentially) to share."[7]
Genetic information, it should be noted, is not only derived from direct
DNA analysis. It can be extracted from examining pedigrees, in which a
type of disease has shown a pattern of inheritance within a family. In fact,
crude pedigrees, or even less technical knowledge, may spur someone on to
donate their DNA for the purpose of scientific research. That is, knowing
that a particular disease is "in the family" may lead an individual to realize
the biological value of donating his or her DNA to science. However, in
this scenario, the information and potential link to a specific disease is suspected prior to donation, as opposed to direct DNA analysis, in which un-
suspected genetic susceptibilities can be unearthed, linking the familial history of disease to a particular gene. For instance, someone might donate
their material because she knows that high blood pressure has historically
plagued her family, which signifies a hereditary risk; however, a researcher
may subsequently reveal a genetic predisposition to Alzheimer's, which is
distant from the intention, and expected results, of donation. These results
would then be included in the individual's DNA profile, expanding the
medical utility of the sample, but potentially compromising the individual's objectives.

The scenario above is one of many potential imbalances between research
objectives and individual expectations. Because research is not the only mo-
tivating force behind establishing and utilizing DNA banks, it is important
to understand how DNA data banks are intended to meet the demands of
business. The answer is inherent in the definition of a data bank. The central
location of samples and data remove many of the variables and administrative obstacles that impede research. Information is on hand, and the necessary effort to find the exact type of samples required for gene discovery or
gene-specific drug development is decreased. Furthermore, the type of data
and samples extracted from the bank for a particular type of research can be
narrowed to include only those that fit specific criteria, dramatically reducing the "noise" that irrelevant samples might contribute.

---

[7]McEwan, Jean. DNA Databanks. In Rothstein, Mark A. (Ed.), *Genetic Secrets: Protecting Privacy and Confidentiality in the Genetic Era,* p. 231. New Haven, CT: Yale University Press,
1997.

Simply put, a family that exhibits a hereditary pattern of disease might hold the key to discovering the genetic basis of the affliction. With enough samples, DNA data banks will contain a significant distribution of relevant material to identify the subpopulation of donors needed for the research program. This isolated population will hopefully possess enough homogeneity to remove most of the genetic variables unassociated with the disease. Because the greatest genetic homogeneity will help in identifying the abnormalities that contribute to a disease or its symptoms, populations of closely related individuals are the best starting point. For instance, deCODE genetics, recipient of many ethical critiques, has based its business strategy on using Iceland's genetically isolated population to begin studies of genetic disease. Within this population, families who exhibit a particular disease may contain a corresponding polymorphism (an aberration in their genetic code) that is passed on from generation to generation. Although a family with such a pattern of inheritance is valuable to the study of disease, the surrounding homogeneous population is equally important because it makes it easier to pinpoint common genetic variations that might be overlooked in the presence of the many different variations exhibited by heterogeneous populations. In other words, the more that the population has in common, the easier it is to identify the differences, especially in polygenic diseases. Polygenic diseases, disorders resulting from multiple genes rather than just one, may work in complex combinations. Isolating a single gene associated with disease pathology may not reveal a consistent genetic pattern because of the effects that other genes have on the condition. Thus, inherited polygenic diseases almost have to be analyzed within a homogenous group to reveal the genes that work in concert to influence disease.

Finding genes associated with diseases is a scientific procedure of increasing specificity. The process begins by identifying those individuals who carry the disease. The next step is locating the chromosome that contains the gene of interest, which is followed by finding the gene itself. After this gene is identified, it is tested for different types of mutations that might correspond to different types of disease. Of course, it is a much more complex process than this description might suggest, but advances in human genetics are simplifying the process every day. The Human Genome Project (HGP) has uncovered a vast amount of information that lay hidden within the human genome's three

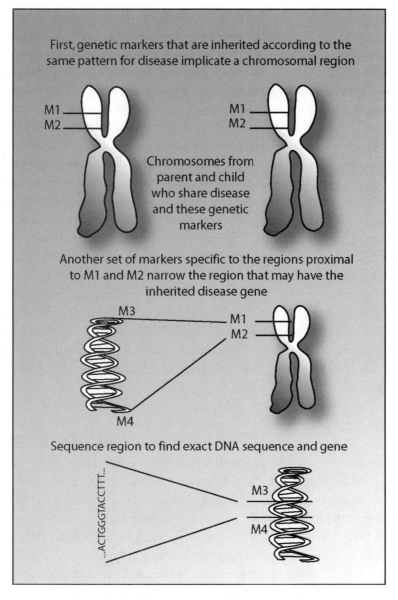

Figure 3-2   Gene mapping with genetic markers.

billion base pairs. Recently, the number of genes estimated to comprise humanity's genetic structure has decreased from 100,000 to 30,000–40,000,[8] although the calculation is far from definitive. Indeed, researchers dispute the number based on various biological bases. Those who argue for fewer claim that humans have more control genes, whereas those who claim more believe that more research has to be performed to validate the number.

Prior to this discovery, the HGP had unearthed a substantial number of genetic markers. These markers are aberrations in otherwise well-conserved genetic sequences. Because the frequency of these aberrations is fairly high, they can be easily identified in different individuals and used as a standard of comparison when necessary. Of the over 6000 markers, about 400 are used in the first stages of identifying a gene locus. These 400 markers are chosen because of their broad distribution across the entire human genome. When particular sets of markers are inherited according to the same pattern of disease, the potential areas for the disease gene are accordingly narrowed. At this stage, a similar experiment is performed using more precise markers that flank these regions. After these markers increase the analytic specificity, enough areas have been eliminated (hopefully all but one) as potential gene sites to allow another, more precise, method of analysis. Finally, regions close to those areas of interest are sequenced to determine the exact DNA sequence and location of the gene (Figure 3-2).

Following gene discovery, polymorphic analysis is the next step. For instance, carefully chosen DNA was used to discover the gene involved in cystic fibrosis, a genetic disorder that affects the respiratory, digestive, and reproductive systems. The goal was to find specific mutations in the *CFTR* gene, and then correlate these differences to disease states. Although the process of genotype–phenotype correlation is difficult for a number of reasons (multiple genes are often involved with disease, or aberrations may occur at many different locations in a gene), it can be enough to understand the gene and the corresponding protein that it creates to either identify a drug target or develop a diagnostic test. In the case of the latter, enough data

---

[8]Ventner, J. Craig, et. al. The Sequence of the Human Genome. Human Genome Special Issue. *Science, 291* 5507, 1304–1351, February 16, 2001. The Genome International Sequencing Consortium. Initial Sequencing and Analysis of the Human Genome. *Nature,* 409, 860–921, February 15, 2001.

needs to be gathered to attribute specific mutations in the gene to different stages of the disease, although some genetic irregularities may remain uncharacterized due to their infrequency.

As indicated above, the population chosen for gene discovery is critical to streamlining the process. Family members, especially siblings, who carry the disease are often the ideal choice; but lacking such a population, there are other parameters that can be used to choose a cohort. No matter which strategy is chosen, the prevailing notion is to limit the pool to as homogeneous a group as possible. The great wish of many scientists is to find genetically homogeneous populations, which is still very difficult in the modern, diasporous world. Because they are so rare, corresponding biological materials are quite valuable—something deCODE Genetics hoped to lucratively obtain when it lobbied the Althing (Iceland's Parliament) for exclusive rights to Iceland's health database and DNA (data) bank. Homogeneous populations are not the only valuable demographic, however. Although similar, isolated populations ease gene discovery, common diseases are also good parameters to distinguish a study population.

Competition in the pharmaceutical industry has been fueled by emerging technology and the high cost of drug development. These factors have played a substantial role in pushing the industry toward the union of genomics and pharmacology known as pharmacogenomics or pharmacogenetics. There are two important benefits to these related fields, one from the industry's and another from the consumer's perspective. In the case of the former, new technologies linking the drug discovery process to genetics will ideally decrease the amount of time involved in developing therapies, while decreasing the costs involved. Thus, data banks offer the biological material and information that would be pivotal to accomplishing this goal. Identifying specific drug targets based on genetics is more rational and efficient than creating a host of potential drugs (often hundreds) and testing each compound individually, betting that one *might* work. Any measure that can hone this process, which often takes from two to ten years, is critical to the future of biotechnology.[9]

---

[9]Ernst & Young. *Convergence: The Biotechnology Industry Report.* Millennium Edition. 2000.

The beneficial implications for consumers are directly linked to this method. When more drugs are created based on the genetics of disease, personalized medicine, that is, medicine delivered based on the precise understanding of an individual's genetic makeup, will herald a new, efficient treatment methodology. Therapeutic regimens will consist of specific drug cocktails tailored to an individual's unique genetic blueprint, which will indicate which conditions to treat and which drugs will work best.

These new scientific demands represent tremendous business opportunities that have bred a new form of company, one dedicated to storing DNA and genetic information. Although pharmaceutical companies and hospitals see the value in this, and thus maintain their own data banks, many of their repositories are built upon clinical trials for a particular treatment that was tested many years ago, or from the excision of material during routine medical procedures; they are not always applicable to current research agendas because they were established for other purposes. However, there are groups that have recognized how linking data banks to another new technology can create the means of ushering in this new age in medicine. Surprisingly, this new technology is not medically based, although it promises to be inseparable from medicine in the future. It is based on the Internet, or more precisely information technology. When it becomes difficult to go to those populations that would be useful to the study of disease, the strategy is to bring that population to you. This viewpoint is the basis of DNA Sciences, an Internet-based DNA data banking company: "We use the Internet's communication power to recruit healthy volunteers and those affected with disease into very large studies on genetics. The goal is to understand all the major genetic variables that contribute to disease. Without the Internet these studies might be very difficult to conduct and take much more time. In this sense, the Internet enables our Big Biology Project, and we are confident that many others will follow."[10] And as predicted by DNA Sciences, a number of companies have emerged to fill this space. Some examples are the already mentioned deCODE and First Genetic Trust, a group which hopes to ex-

---

[10]Reinhoff, Hugh Y. and Clark, Jim. The Convergence of Biotechnology and IT: Genetics as a Model. In Ernst & Young, *Convergence: The Biotechnology Industry Report*. Millennium Edition. 2000, pg. 38.

pand DNA data banking beyond just the business-to-business market and into the business-to-consumer realm. Not only will the company use the Internet to recruit and stay in touch with donors, the participants may use First Genetic Trust's Internet-based personal profile information for customized health care.[11] Also, some patient advocacy groups for specific diseases are creating their own data banks for discovery purposes; both the Epilepsy Foundation of America (EFA) and the Pseudoxanthoma Elasticum (PXE) Support Group have begun such initiatives.

III. The Benefits

If companies succeed with strict commercialization of DNA data banking, that is, data banking as its own industry, the potential to identify the genetic basis of and treat disease is immense. Similar benefits will be realized by biotechnology and pharmaceutical firms that already maintain their own stores of DNA profiles and human tissue; however, the range of applications and administrative freedom associated with outsourcing this task allots more time and resources for the drug discovery and development process. Previous efforts to use tissue banks (prior to the advent of techniques that simplified DNA analysis) have a rich history of success. Tissue samples have been critical in understanding diseases ranging from atherosclerosis to the carcinogenic effects of previously, commonly prescribed treatments during pregnancy. Working at the level of DNA, these types of discoveries can increase in their range and depth, moving beyond correlations, establishing diagnostic value and, ultimately remedies.

The great benefit of DNA data banks and tissue banks lies in how they ease and streamline the development of therapeutics. One might wonder what compels individuals to contribute their material to DNA Sciences or one of its counterparts. Often, a sense of altruism is the motivator. This is not unheard of, and a great example of civic pride inspiring massive contri-

---

[11]Fikes, Bradley J. First Genetic Trust Reaches out to Patients . . . but Can it Earn Trust, Clinical Samples, and a Profit? *http://www.doubletwist.com/news/columns,* October 27, 2000, p. 1.

butions to medical progress occurs frequently, as in blood drives. In a recent study in France, Paul Rabinow, Chair of Anthropology at UC Berkeley, explains the level of responsibility that individuals feel toward their community when inspired to help medical progress. He further explains how the tradition of donating blood conferred cultural credence to other initiatives that involve tissue donation.[12]

In the case of disease, direct benefits to the donor might eclipse the notion of civic pride, although, most likely, both work in tandem. Those who donate materials are carriers, victims, or close to someone who suffers from disease. These people, who witness the pattern of disease inheritance in their families, are likely to be the first to recognize the benefits that may result based on the analysis of their genes. Drugs and therapies developed for a particular disease will benefit those people on whose biology the treatment is based. In a way, donating material is a means of helping oneself, or future generations of one's family.

Furthermore, for those who value the concept of choice in reproduction, the concept of learning more about the genetic nature of disease offers greater insight into reproductive planning, and the range of alternatives will increase as more information is gained regarding inherited disorders. Although it may seem that this is an argument that tips the scale toward one end of the abortion debate (a debate engaged in Chapter 5), it favors neither side. Information can be gleaned that might seem the reason for choosing to terminate a pregnancy; however, the goal of medicine, both socially and economically, is to develop treatments. If a therapeutic drag can be delivered in the womb following a diagnosis of disease, many individuals will be able to exclude the possibility of abortion. Granted, this is a point that arises after a cure is developed and approved. In light of this point, it should be noted that due to the nature of biotechnology, a genetic test is developed prior to treatments; indeed, there are not many genetically based remedies on the market at the time of this writing. However, the promise of biotechnology and the here and now are on a path toward convergence, and upon this meeting the ever-expanding choices facing individuals will only increase.

---

[12]Rabinow, Paul. *French DNA: Trouble in Purgatory.* Chicago: University of Chicago Press, 1999.

Perhaps the greatest benefit, even beyond diagnostic genetic testing (which is discussed in Chapter 4), is the new generation of gene-targeted or protein-targeted drugs, mentioned briefly above. The new field of pharmacogenomics promises to deliver personalized medicine. Envision a day when a small chip, or microarray, no larger than a microscope slide contains a number of DNA molecules that correspond to commonly known disease-causing genes. A doctor can take a swab and wipe the inside of the patient's cheek; then the physician places the swab in a machine that extracts and amplifies the DNA contained in that sample. Next, the amplified material is transferred to the chip, and after a few minutes, the test indicates the presence of disease, its status, as well as any predispositions to a disorder or series of disorders (Figure 3-3). Because of the patient's genetic individuality, a unique set of therapies will yield the best results, and the corresponding prescription is printed out after the incredibly powerful bioinformatic computer correlates the exact genetic nature of disease to an exact suite of remedies.

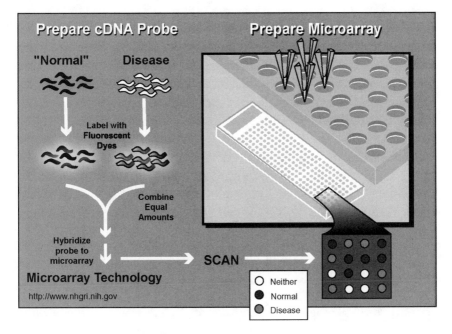

Figure 3-3 Microarray technology.

As science discovers the genetic basis of diseases, cures, and patterns of drug response, an individual may receive therapeutic updates as the DNA data bank is scanned during the simultaneous process of updating files and diagnosing patients based on stored information. However, this type of caregiving will only become possible when enough genetic information is gathered from enough individuals, and it carries a complimentary set of ethical concerns, which will be addressed in Chapter 4.

Before delving into the ethical issues that surround DNA data banking, it is important to note that there are more applications and benefits than are addressed in this book. Because the focus of this work is the business of biotechnology, its emphasis is limited to corporate research goals: gene discovery, predicting disease, developing gene-based diagnostics, and treating genetic disorders. However, there are other purposes that are in use or will be used and deserve mention. In many states, convicted criminals (the conditions used to choose what types of crimes warrant data banking vary from state to state) must contribute samples to a DNA bank, taken for identification in case of recidivism. The largest single data bank in the United States is held by the U.S. military to identify the remains of unidentifiable, deceased soldiers via DNA fingerprinting. Furthermore, some parents use data banks to store their children's genetic information and material in the sad case that identification is necessary at some point in the future. Although these are all important examples of data banking, some of whose ethical concerns overlap with those of biotechnology, the following pages will be dedicated to specific, corporate, biotechnological applications of the technology.

IV. The Issues

Although the benefits of DNA data banking may seem both exciting and substantial, the technology is the subject of great bioethical scrutiny. As technology progresses on both the biotechnological and information technology fronts, it will only become easier and more attractive to maintain or create such storehouses. Increasingly, individuals will be recruited to contribute to medical progress by simply providing some blood or other form of tissue that can yield DNA. In giving up that DNA, individuals will also be giving up information about themselves and their families, which may be beyond their self-understanding. Information about their past, present, and future may be systematically revealed to others and stored in comprehensive profiles, prior to the donor knowing about it, if at all.

Most likely, the few lines mentioned above already indicate the types of ethical issues that surround DNA data banking. In a very concise way, George Annas, a bioethicist and Chairman of the Health Law Department at the Boston University School of Public Health, has developed a term to describe the unique status of information revealed by genetic analysis: "DNA future diary."[13] Although this term is problematic because it may emphasize genetic reductionism (it should be noted that Annas does not promote a deterministic definition, but rather its use as a metaphor specifically relating to privacy issues), it is useful as a heuristic tool to understand the issues surrounding genetic technologies. In his description of genetic information, Annas states:

> Increasingly, precise genetic information has the potentital to radically alter our life choices because control and access to the information contained in our individual genes give others potential power over our personal lives by providing a basis not only for counseling, but also for stigmatization and discrimination. . . . It is in a real sense a future diary (although a probabilistic one), and it is written in a code that we have not yet cracked. But the code is being broken piece by piece, such that holders of a sample of any individual's DNA will be able to learn more and more about that individual and his or her family in the future as the code is broken.[14]

Beyond this critique is the effect of information technology, which increases general accessibility to data. All of these concerns speak to the major concern that ethicists have raised against DNA data banking: privacy.

---

[13]Annas, George J. Privacy Rules for DNA Databanks. Protecting Coded "Future Diaries." *JAMA, 270* (19), 2346–2350, 1993.
[14]Annas, George J. *Some Choice: Law Medicine and the Market.* New York: Oxford University Press, 1998, pp. 98–100.

Privacy and confidentiality in biotechnology have received as much attention as data banking itself. Although these are very important issues to the technology, they are not rooted in data banking. Health records have always been subject to a privacy debate, although the intensity has increased alongside the emergence of genetic technologies. The unease associated with relinquishing information to another is exemplified in a passage by Nobelist Alexander Solzhenitsyn:

> As every man goes through life he fills in a number of forms for the record. . . . A man's answer to one question on one form becomes a little thread, permanently connecting him to the local center of personnel records administration. There are thus hundreds of little threads radiating from every man. . . . They are not visible, they are not material, but every man is constantly aware of their existence. . . . Each man, permanently aware of his own invisible threads, naturally develops a respect for the people who manipulate the threads . . . and for these people's authority.[15]

Before continuing the discussion of privacy and confidentiality, it is important to define these terms as they relate to bioethics. Privacy has many applications, but the most relevant explanations are offered by some classic studies on the topic. The first, published in the *Harvard Law Review* in 1890 by Samuel D. Warren and Louis D. Brandeis, defines privacy as "the right to be let alone."[16] Political scientist Alan Westin furthered the notion by stating in his book *Privacy and Freedom* that privacy is the "claim of individuals, groups, or institutions to determine for themselves when, how and to what extent information about them is communicated to others."[17] Although privacy deals with control of information, confidentiality is an expansion on Westin's definition. In the case of genetic technologies, a degree of privacy must be relinquished whenever the information is shared; howev-

---

[15]Solzhenitsyn, Alexander. *Cancer Ward.* New York: Farrar, Strauss & Giroux, 1969. Quoted in Annas, *Some Choice,* 1998.
[16]Warren, Samuel D. and Brandeis, Louis D. The Right to Privacy. *Harvard Law Review, 4,* 193, 1890.
[17]Westin, Alan. *Privacy and Freedom.* New York: Atheneum, 1967, p. 7.

er, this concession is usually based on mutual trust between the parties involved. Confidentiality, then, is honoring the expectation that privacy will be respected and any flow of information will be restricted to only those people who exhibit and understand that trust.

Bioethicists have approached the privacy debate by dissecting the interpretation of privacy into four types. These divisions are: physical, informational, decisional, and proprietary. Each type of privacy has a particular application with respect to DNA banking and data banking.[18]

Physical privacy is often irrelevant in the data banking debate. Relating to the invasion of one's body, a transgression that violated physical privacy would be extracting human tissue without the consent of the individual. In the United States, violating this aspect of privacy is illegal, and in the case of data banking as a business, it is almost an irrelevant point. The recruitment process includes a number of forms and waivers that need to be signed by a donor. When constructing a data bank, the individuals targeted have very specific genetic histories; thus, tools such as the Internet are used to announce that the company needs a specific type of donor. The donor is usually well aware that phlebotomy is a necessary precursor to isolating DNA. Furthermore, the minimum legal requirements for participation in research require a consent form that highlights the risks and benefits associated with the experiment, which includes the negligible side effects of having blood drawn.[19]

Perhaps the most important type of privacy in the DNA banking and data banking debate is informational. Informational privacy relates to the extensive amount of information that can be revealed in a future diary. More importantly, informational privacy focuses on third party accessibility to this data. Concerns surrounding this aspect of privacy ask the simple question: Who should have access to the information? The relevant follow-up ques-

[18]Allen, Anita. Genetic Privacy: Emerging Concepts and Values. In Rothstein, Mark (Ed.), *Genetic Secrets: Protecting Privacy and Confidentiality.* New Haven, CT: Yale University Press, 1997. Greely, Henry T. Breaking the Stalemate: A Prospective Regulatory Framework for Unforeseen Research Uses of Human Tissue Samples and Health Information. *Wake Forest Law Review, 34,* 737–766. 1999. Annas, George J. *Some Choice,* 1998. Robertson, John A. Privacy Issues in Second Stage Genomics. *Jurimetrics, 40,* 59–76, 1999.
[19]The Common Rule. 45 C.F.R. pt 46 (1998). The Food and Drug Administration has a its own version of the Common Rule—21 C.F.R. pt 50 (1998).

tions are equally simplistic: Why and how? The straightforward answer may seem to be that the only individuals who should have access to the information are the person who donated the material and the investigator performing the research, which would suit most people just fine. However, this solution is impractical. When research is performed, data is entered into records for comparison's sake, and there are a number of people who have access to these records. Moreover, an argument can be made that if health-related information is revealed, the donor's caregiver ought to be informed as well. Furthermore, there is the unique familial characteristic of genetic information; kin share genetics, thus the information may bear upon the relatives of the donor. Whether these individuals ought to have access to the information is an equally valid question; yet, how does one balance this with "the right to be let alone"? Conclusive information for the donor may be speculative for the donor's family members because the relatives have not been tested; however, the moral dilemma as to what conditions should be satisfied for the individual to notify relatives is amplified by the same dilemma facing the investigator. Is either person obligated to inform someone with an unknown likelihood of disease just because the chances are calculably higher than the rest of the population? More precisely, at what level of certainty, if at all, should an investigator inform someone of a disease they *might* have? What if there is no cure, or the disease is not grave?

Although the issues mentioned thus far are very important, they arise as a result of trying to both convey the benefits of research and to improve the health status of different individuals. Because of the motivation behind those who would want the details, third-party access to information is perhaps the central issue with respect to data banking. The most insidious problems lie in the realm of insurance and employment, or, more appropriately, genetic discrimination. If information is released to a donor's medical records, then it is safe to assume that insurers can access the information, and most probably will. Similar concerns emerge with respect to employers gaining access to the information.

In the case of insurance, policies are contracted through a process of underwriting risk. The greater an individual's risk of contracting an expensive disease, then either the higher the premium or the greater the chance of being refused a policy. Although there is evidence that insurance policies have been denied on the basis of genetic predispositions toward disease,[20] the magnitude of this issue is often downplayed by the prevailing notion of group insurance. That is, most people in the United States are covered by group policies that are provided by employers; however, this interpretation offers false security for two reasons. 1) The issue should not be obscured because evidence suggests that the problem is smaller in scale than the most critical assessments; if denying insurance is an issue, it is an issue regardless of the small number of people affected. 2) If employers gain access to the same type of information that insurers do, then they can discriminate based on the information; healthcare costs that increase the group insurance premium can influence the decision to hire or fire an individual. Along with the burden of not finding healthcare, there is the added injury of the loss or denial of employment.

It is not the goal of this text to debate whether insurance is a right or a privilege. The complications of this specific insurance debate do include ethical arguments; however, the topic deserves more attention than can be afforded in this text. The point is that engaging all stakeholders is critical to good business practices, and thus all serious debates surrounding ethics require attention. Whichever side a company chooses in this policy struggle, it should do so while understanding the opposing arguments, allowing the most rational choice.

Genetic discrimination in the workplace or insurance may seem as if it is more an issue for the courts than corporate policy, but its severity is grand enough to invite attention before the issue ever reaches the courtroom; an individual may be irreparably wronged before the case even reaches the court's docket. Furthermore, should it go that far, companies stand to lose quite a bit. In 2001, the Equal Employment Opportunity Commission sued the Burlington Northern Santa Fe (BNSF) Railway Corporation because the latter used genetic testing to identify susceptibility to carpal tunnel syndrome.[21] The case resulted in a settlement, upon which one condition was:

---

[20]Lapham, Virginia E. Kozma, Chahira, and Weiss, Joan O. Genetic Discrimination: Perspectives of Consumers. *Science, 274* (5287), 621–624, 1996.
[21]Commission Sues Railroad to End Genetic Testing in Work Injury Cases. *New York Times,* February 10, 2001.

"In its ongoing investigation of the initial charge filings, EEOC may seek compensatory and punitive damages up to $300,000 per individual (the statutory cap) for a class of claimants ranging from 20 to 30 BNSF workers who were either subjected to genetic testing or retaliated against for failing to submit to such tests."[22] Even though BNSF had discontinued its genetic testing practices, it had done so too late to save itself from the damage to its image, and the pending court settlement payments levied against them.

Because of cases similar to that of the EEOC and BNSF, there might also be the temptation to push the issue off of the data banking industry and solely onto the companies who *make* these decisions, since data banking only *informs* these decisions. However, disregard for the donor's well-being is a surefire way to guarantee that people will not contribute their tissue to a storage facility. If the risks are not addressed and mitigated, the potential benefits are not compelling in the light of the potential problems.

Informational/accessibility issues extend beyond the tangible effects of insurance and employment repercussions. When new information is revealed, donors may not want anyone to know about the data, regardless of whether there are actual repercussions. At the heart of the debate over how to confront these issues lie two different ethical approaches: consequentialism and deontology. Those that result in actual, observable consequences like denied insurance are consequentialist. In its simplest form, consequentialism distills moral questions into a discussion over actual harm. The consequences of an action determine whether or not the action was right or wrong. If no harm is felt, then the action is ethical, but if there is some type of damage, then the opposite is true. At one end of the continuum there is no violation and no harm and at the other end there is both violation and harm. At the center of this continuum is violation and no harm. The questions that matter in this mode of thought are at one extreme where violations lead to harm, as the other conditions do not compromise the subjects in any observable way.

Deontological harm is also characterized as dignatory harm. In traditional philosophy, this position is best known in reference to Immanuel Kant. Kant contends that individuals have an interest in being treated with a de-

---

[22]EEOC Settles ADA Suit against BNSF for Genetic Bias. *http://www.eeoc.gov/press/4-18-01.html.*

gree of dignity and respect based on a common understanding of humanity. Often simplified into the Golden Rule of "do unto others as you would have them do unto you," the concern focuses on general conceptions of moral conduct; these considerations manifest themselves in an agreement over confidentiality in data banking. A donor expects that his or her trust will not be violated. Although it may not seem like a major violation if an unsanctioned individual "harmlessly" accesses confidential data, the fact that trust is compromised is still deontologically suspect. Again, the practical impact might appear negligible, but from a business perspective, the question arises: If those who ought not have access to this sensitive information can and do retrieve these files, is there a breakdown of trust in the organization that compromises more than just its corporate values?

Third-party accessibility becomes more problematic in the connected world. With the Internet and intranets, the number of eyes that can view information increases. Many of the databases that store information have general access codes that allow total access or none at all. When most employees of a data bank do not need access to information yet can get it, the dignatory harms associated with data banking become as much a matter of technology as morality. Collaborations further complicate this matter, as do mergers and acquisitions. With regard to many of these cases, Greg Sabatino, information technology specialist and cofounder of netNumina Solutions states:

> Information structures often suffer when companies are taken over and merged. Emphasis is often placed as much on acquiring skills, know-how, and relationships as much as it is on data. Due to the level of effort required, the integration of raw database tables may be the last undertaking in these scenarios. When they are, limiting accessibility might be quite the opposite of the intended goals.

The solution may lie in policy, but most people would feel safer if technological barriers were put in place as primary, rather than secondary measures.

Another area of concern regarding informational privacy lies with the legal, forced disclosure of information. Although insurance agencies and employers may access data, they can only do so when the information is placed in a record to which they are allowed access. However, because DNA data banks are very convenient sources of forensic and paternity data, it is not unreasonable to assume that, as data banks and DNA evidence gain greater prominence, courts may subpoena samples to support or dispel evidence. Of course, the counterpoint can be raised: Why should we systematically deny access to evidence that might serve justice. No one wants to help a criminal escape, or help a "deadbeat dad" avoid child support payments. Although the information might help in criminal or paternity cases, it is a means of using one's body against them, without the defensive benefit of the Fifth Amendment. Despite this heavy question, the violation of privacy still remains a formidable ethical concern, and to compromise it does so at the level of human rights. Violating one person's rights on this debatable foundation may set a precedent for conduct that can prove damaging if performed unreflexively. Regardless of where a business falls on this debate (both sides have powerful ethical arguments), the tension between social justice and individual rights might be palpable enough to discourage donor participation.

The third type of privacy, decisional, focuses on the actual use of data or tissue. These institutions are termed banks because they store materials that are used both for the present and the future. In the immediate sense, the donor is usually made fully aware of the intended use of the material. Indeed, the *informed* part of Informed Consent is meant to disclose the anticipated use of the material, as well as the associated risks. Most donors contribute their material because they want to see a particular disorder cured, thus the act of donation is an affirmation of the research being performed. However, future use may touch upon research that the donor opposes. For instance, someone might donate their material for breast and ovarian cancer research; however, the data may be appropriated for studies involving the incidence of cancer in Ashkenazi Jews, which they oppose due to concerns over the stigmatization of that group. Despite this ideological position, the informed consent process ought not bring into conflict the conditions of donation (what the individual understood the limits of his or her participation to be) and the objective goals of science, despite the value of the latter.

This example speaks to another concern that fits loosely under decisional privacy: the impact of research on groups. In many instances, particular demographics are chosen due to the prevalence of a particular disease within that group. This assemblage of people may be distinguished by race or gender, and donors might justifiably fear that the experimental results gained may have a negative, social consequence. Stigmatization or furthering stereotypes might be extrapolated from the research conclusions, or the characterization of the experimental cohort. Such problems have historical foundation such as with sickle cell anemia for the African-American population in the United States. In the 1970s, the United States Air Force refused African Americans who tested positive for carrying an allele causing sickle cell anemia the opportunity to serve as pilots for what most agree to be a racially motivated decision. Genetics in this case were either misunderstood or manipulated for a dubious end. The presence of one allele in this case did not impart the symptoms of the disease and, in fact, it represents a benefit in some geographic areas where a single "disease" allele confers resistance to malaria. Whether genetic information is a valid basis for justifying these types of racially motivated decisions is not at issue, that it might be used as justification is.

Proprietary privacy enters as the final type in this discussion. Proprietary privacy refers to the ownership of material and the distribution of rewards based on the research use of biological samples. One of the most famous court cases in biotech sets the stage for this notion: *John Moore v. the Regents of the University of California.* In this case, John Moore received treatment at a teaching hospital affiliated with the University of California, Los Angeles. Prior to a procedure that resulted in removing his spleen, Moore signed the usual consent and waiver forms, without realizing that these forms allowed the retention of tissue for research purposes. Although Moore should have paid more attention to the forms that he signed, particularly at a research hospital, it was never fully explained that his tissue was extracted for its potential commercial use, as well as health and research purposes. Unaware that his biological material was being collected to develop a very valuable cell line, the constant requests to return to UCLA for more testing aroused his suspicion. Moore contacted a lawyer following repeated attempts by the investigator to have him sign a new consent form. The new cell line, named for Moore as the Mo Cell Line, was eventually patented and proved to be very valuable, and Moore wanted a financial stake in the research applications. Although there might be an intuitive feeling that individuals have a right to the benefits realized through the exploitation of their tissue, the courts saw it differently. With respect to Moore's claim that his "property" had been illegally converted for another's profit (legally termed conversion), the court took the side of general scientific research: Moore's claim "threaten[ed] with disabling civil liability innocent parties who are engaged in socially useful activities, such as researchers who have no reason to believe that their use of a particular cell sample is, or may be, against a donor's wishes."[23]

This legal decision is an important victory for biotechnology, affirming the value of the industry's research, but it does not constitute the final ethical say. That there was so much controversy associated with this case only highlights the need to address issues regarding proprietary privacy. Many people disagree with this decision, and persons on both sides of this case agree that the failure of the consent process to articulate the potential commercial interests of the investigator is an inexcusable failing of procedure. Furthermore, the implication that biotechnology companies can own genetic material becomes problematic in places where there are underprivileged, poverty-stricken, or indigenous populations who do not understand the implications of technology. In many of these cases, there is a high level of genetic homology. This is as much an issue domestically as it is in developing countries, and it may even be worse in foreign lands. A genetically conserved populace may contain very valuable DNA, and foreign companies have more leverage and knowledge to abuse, consciously or not, at the expense of an impoverished and/or uneducated population. There seems to be no clear-cut solution to this issue, but to adequately address the ethical dimension of this debate in the future, the interests of all stakeholders need to be regarded and carefully considered.

V. Industry

In a poll performed in 1995, 148 diagnostic laboratories were surveyed about their intentions regarding DNA banks. Of this sample, 90% of these groups had begun to bring DNA banks and data banks in-house. Further-

---

[23] *John Moore v. University of California Regents.* 793 P.2d 493. 1990.

more, one-third of the companies indicated that they planned on expanding their banks toward a service-based model over the next few years.[24] Today, in addition to established companies that are expanding labs, companies based solely on DNA and DNA data banking are emerging. The business opportunity for data banking breaks down into two types of services: 1) providing information or samples to companies or universities that need particular types of DNA samples for their research and development or performing the research in-house, and 2) storing DNA for individuals who wish to maintain a central location of their genetic material and data.

Although there are two corporate approaches to data banking, a business-to-business or business-to-consumer model, few companies dedicated to data banking will settle for one over the other because the infrastructure for maintaining either form is easily adapted to the other. Whether a group is prepared to address both as it launches its data bank is a different question, but it is not unreasonable to expect that a company dedicated to providing samples to internal or external research programs will also provide individual banking service to its donors. Because of this probability, all of the ethical considerations mentioned in the previous section are issues that *all* prospective and current DNA data bankers ought to consider.

To understand which steps banking companies can take to address the ethical concerns facing the industry, it is useful to see what some businesses have already done. Bioethics is an important part of this industry, whether apparent or not, simply by virtue of some of the laws that govern research regarding human subjects.[25] Although there is legal precedent for including bioethics in corporate operations, heeding the law and truly adopting bioethical measures are significantly different. To explain the bioethical measures that corporations have applied, the data banking company DNA Sciences, Inc. will be used as an example. Although the firm has placed a tremendous emphasis on privacy, it is an interesting case because the corpo-

---

[24]McEwen, Jean and Reilly, Philip R. A Survey of DNA Diagnostic Laboratories Regarding DNA Banking. *American Journal of Human Genetics, 56,* 1477–1486, 1995.
[25]Jonsen, Albert R. *The Birth of Bioethics.* New York: Oxford University Press, 1998, 342–344, 355–356. The laws that protect human research subjects (defined as any human being or biological material derived from human beings that can be traced back to its source) are deeply rooted in bioethical debates.

ration bet that the future of this industry depends on a platform with its own privacy concerns: the Internet.

In many ways, the corporate goals of DNA Sciences are very similar to most genomics-based biotech companies. According to the company, "DNA Sciences is a genetics discovery company focused on identifying the genetic basis of disease susceptibility, disease progression and response to drug treatment."[26] As indicated throughout this discussion, to meet their goals, the company requires many different DNA samples. As they collect more and more, their information stores will increase as well, until they are capable of doing all that they promise. What distinguishes this group from traditional data banks is their emphasis on the Internet. By using the World Wide Web, DNA Sciences can recruit participants, as well as offer information services that give back to the donors. "Now, consumers can learn about genetics and, ultimately, take control of their own genetic information in a safe and secure locale. Absent any legislative safeguards prohibiting the misuse of genetic information, consumers—especially health consumers—will need a discrete service that can help them identify and manage genetically identified disease risks."[27] More than a recruitment tool and a means of maintaining contact with donors, DNA Sciences also uses the Internet as an educational tool, hosting chat rooms and online events that discuss genetics, disease, and the ethics of biotechnology.

To participate in these services, an individual must fill out the requisite forms and join the site. This process begins the data acquisition phase of the company's operations. Although there has been no donation of biological material, information is being accumulated to help the company increase its statistical data vis-à-vis disease. Aware of the ethical implications with regard to information and privacy, DNA Sciences will not allow participation by anyone who does not read and agree to their policies. Although many Internet-based companies receive the common criticism that the privacy policy is too complex and detailed to draw anyone's true attention, DNA Sciences makes a noticeable effort to simplify the process. General tenets are put forth,[28] while more detail is available should anyone require it by clicking on individual

[26]*http://www.dna.com.*
[27]Reinhoff and Clark. *Convergence: The Biotechnology Industry Report.* Millennium Edition. 2000, p. 38.
[28]*www.dna.com.*

points, including an e-mail address that is answered by the company's Director of Public Policy and Bioethics. Although these points are important concerns that implicate medical information, they are issues related to the Internet services offered by the company, regardless of whether biological material is donated to the company or not.

If an individual decides to donate material to the company, he or she enters into a new level of involvement with DNA Sciences. Furthermore, the part of the company that deals with biological material is regarded as its own entity, the Gene Trust, an appropriate choice of words—there are increased measures of protection as soon as someone decides to join the Gene Trust. Dr. Joseph W. H. Lough, DNA Sciences' Director of Public Policy and Bioethics states that:

> When visitors decide they want to explore participating in the Gene Trust, we offer three additional layers of security.
>
> First, because participants will provide us with personal and family health information, we include this information in the kinds of information we will not share with third parties without participants' express, written permission.
>
> Second, because Gene Trust participants become research participants as defined by Federal Code, we let participants know exactly what they can expect as a research participant, how they can contact us, and what further steps they can take in order to contribute their DNA to our research.
>
> Third, when you decide to have your blood drawn and agree to participate in our research, DNA Sciences will direct you to our full "Consent Form," which has been reviewed and approved by an independent Institutional Review Board or "IRB," for short. You will then be encouraged to phone a Gene Trust expert who can answer any questions you might have about your participation in our research.

In the first point, the issue of privacy is addressed. By limiting access to information, the donor can be assured that only those parties explicitly chosen by the donor or agreed upon during the consent process will have access to genetic information. Furthermore, DNA Sciences uses coded identifiers to limit the number of employees and researchers who can link information to particular individuals. This practice is standard procedure with many banking facilities, and it is one step from total protection, which is unlinking personal information from the biological material. Unlinking varies from coding, also known as anonymization, in that unlinked information has no translation key. Thus, it is impossible to link the material to an individual by connecting the information that the bank owns. In the case of DNA Sciences, the coding process works as follows: personal family and health information is collected when anyone participates in the services offered by the website (e.g., chat rooms); if someone does not participate in the website services and hopes to join just the Gene Trust, this first stage of information is still collected. Upon donating material, the information is then linked to that sample, but personal identities are replaced by codes. Only a select few people have access to these codes, and release of any information that allows connecting data to individuals is expressly prohibited. According to DNA Sciences' Consent form: "We will perform genetic analysis of your blood sample . . . and we will analyze how these results relate to your Family and Health Information. We will combine the results from many individuals to discover relationships between genes and health. We will not provide you or your health care professional with the results of genetic tests that could be used in making health care or family planning decisions."[29] With this statement, DNA Sciences addresses numerous bioethical issues as they relate to privacy. Specifically, the company will not release data to those parties most likely to relay information to those groups who might use the information for discriminatory purposes, including the donor.

One might ask, why not tell the donor about this information? Although it may seem appropriate to share information with the person it directly concerns, there are some reasons that eclipse this expectation. First, in the case of research, the data may not be conclusive. Without a degree of certainty, the information may not be considered useful to anyone but the research team. This level of protection is only one of many. It is not uncommon for employers to ask and for insurers to demand information about an individual's health. In the case of the former, it may be illegal to make hiring

---

[29]DNA Sciences, Research Subject Information and Consent Form. Approved June 20, 2000. p. 3.

decisions based on health history, but it is easier to allow that information to sway opinion when making firing decisions. In the case of the latter, insurance policies routinely require information regarding family histories and predisposition to disease. If the applicant is aware of a condition or potential condition, revealing the information is necessary and may lead to denial of coverage. To withhold this information might negate the contract at the very least, and result in litigation at the very worst. Finally, the question of revealing information to blood relatives is obviated. Because genetic information may directly implicate the health of a donor's family, whether information ought to be revealed to these relatives is a difficult decision to make. Furthermore, when the information may be inconclusive, and/or when the information may cause mental anguish to the donor, how to handle one's family is hard enough, let alone sharing bad news that extends to them.

Careful examination of DNA Sciences' confidentiality and security measures reveals the extent to which the company is willing to protect a donor's privacy. The policy explains that the following steps are taken to safeguard Gene Trust members:

- A special code number will be assigned to you. Only this code number, and not your personally identifiable information, will appear on your Family and Health Information, your blood sample, and the DNA extracted from it.
- DNA Sciences will keep the key that links this code number to your personally identifiable information under very tight security. DNA Sciences will not give, sell, or allow the key to be used by any unrelated third party.
- Neither your personally identifiable information nor the key will be available to DNA Sciences employees who do not need to use the personally identifiable information for their work.
- All Family and Health Information and Genetic Information will be stored in locked files and on separate computer servers, and only employees who need the information for their work at DNA Sciences will have access to them.[30]

---

[30]Ibid., p. 5.

Coding has already been explained, but the final two points deserve attention in light of the special conditions that apply to them. The Informed Consent form states that the Food and Drug Administration (FDA) and the sponsor of the research will receive information regarding the study. The sponsor refers to any additional groups that are collaborating with DNA Sciences, or who are funding DNA Sciences for a particular type of research. "Research records which may identify you and the Consent form signed by you may be looked at or copied for research and regulatory purposes by the sponsor and may be looked at and/or copied for the regulatory purposes by the Western Institutional Review Board (WIRB)."[31] Although the law requires that certain information be presented to both the FDA and an IRB, revealing information to the sponsor is not. It is with this statement that the balance between research, bioethics, and profit are at their most important crossroad. In this type of research, information is of tremendous value, from both a data and financial perspective. To capitalize on it, a compromise regarding confidentiality may have to be met, and the interest of privacy may suffer. Because of the legal requirements of information disclosure, this discussion will do best to bracket the FDA and IRB, which exist to protect the donor; instead, it should focus solely on the "sponsor." "Absolute confidentiality cannot be guaranteed because of the need to give information to these parties."[32] Although there is a consequentialist argument in favor of revealing identifiable information to the sponsor if it is unlikely to result in repercussions, the contention that dignatory harms may result is also relevant here. The wishes of the donor need to be respected, and although it may seem at first glance that they are not, DNA Sciences promises that they will be. It is important to understand that any company has to set certain policies in place to function, and these policies have to reflect the firm's business goals, as well as its core values. To be honest about its operations and how they affect the potential donor, DNA Sciences discloses the relevant features of its policy in the Consent Form. In this vein, DNA Sciences explains that its business goals *might* require disclosing sensitive information to certain parties. By doing so, they leave the decision in the hands of the potential donor, which is important to their core values; the poli-

---

[31]DNA Sciences, Research Subject Information Consent Form. Approved June, 20, 2000, p. 3.
[32]Ibid.

cies are presented for review, and thus the power of choice is placed in the hands of the potential Gene Trust member *prior* to participation. If this degree of sharing is acceptable to the prospective member, then he or she may join; if not, then there are many other institutions and different ways to participate in research should he or she decide to do so.

The final protective measure regards the placement of biological material and information. In the past, files needed to be kept under lock and key in a physical space to prevent unauthorized access, but the advent of computer networks has complicated the issue of the safe storage of data. Although physical copies continue to require these safeguards, digital data requires more attention. Computer technology has significantly increased the ease by which information is exchanged. The amount of time saved by information networking has made the use of these technologies more than a luxury; they are a necessity to exist and compete in modern technological markets, be they biotechnology or any other. Although most types of businesses are not compromised by these technological measures, data banks face a new dimension in which privacy can be invaded. When genetic information is stored in areas that are routinely used for accounting, inventory, and other operations, the potential for crossover is alarming. At the very least, password protections and other means of limiting access are necessary. Taking protections one step further, DNA Sciences centralizes the information on computer servers that are not accessible by the company's general network. Both physically and connectively isolated, this effort offers an added level of protection for the donor, as DNA Sciences' founders Hugh Reinhoff and Jim Clark explain: "All of these technologies invoke the need to protect individuality, protect confidentiality, and celebrate the sanctity of life."[33]

DNA Sciences provided an excellent case study for this analysis; however, those familiar with this debate may have expected that deCODE genetics would have played a more prominent role in this discussion. There is an important reason for its exclusion. There are many lessons to be learned from deCODE, and the most relevant have already been explored in this chapter, but they are not done so in reference to this company because it does not recruit donors as part of its business operations. Unlike the data banks consid-

---

[33]Reinhoff and Clark. *Convergence: The Biotechnology Industry Report.* Millennium Edition. 2000, p. 39.

ered in this chapter, deCODE has a governmental blessing to its research, and included in that blessing is carte blanche access to Iceland's government-sponsored health information. Very few institutions, if any at all, can be established based on this model, and a political component is integral to its uniqueness:

> Iceland is a small country with a Parliamentarian power structure. The nature of the political system requires a legislative majority as a prerequisite for the formation of a Government; therefore, the Government controls the Legislative Branch as well as the Executive Branch. If deCODE or anyone else for that matter, sells an idea to the Government, the legislative process becomes more or less a formality. In addition, the small size of the society makes the authorities extremely susceptible to lobbying through an integrated network of power typical in smaller societies.[34]

These are not the conditions upon which most data banks will be founded; therefore, the case offers little insight into our discussion. More precisely, ethical concerns primarily affect patient recruitment if a company does not protect those stakeholders who participate. DeCode assumes a different status because participation is presumed, and recruitment is unnecessary in light of political support. Furthermore, a healthcare, labor, or insurance system that does not judge individuals based on their genetic predispositions, as is the case in Iceland, cannot deliver the majority of harms that would preclude a prospective donor from participating. The business will not suffer from poor ethics; although, to be fair to deCode, the company has spent a good deal of time, more than most, in fact, in ethical deliberation. Nonetheless, this model is far from prevalent, which explains its exclusion as a specific example.

VI. Recommendations

Although the response of DNA Sciences to this debate is illustrative, it is just one mode of addressing the issues facing the industry. Each company in

---

[34]Jonatansson, H. *Am. J. of Law and Medicine, 26,* 1200, 38–39

this or similar space must make its own choices as to how it wants to approach the ethics of its research, and the following recommendations are meant to offer a menu of choices, rather than a decree as to the only way to conduct business. All companies contain different cultures and different core values; thus, these specific tools are applicable only to the extent that a firm can actually fit them into its operations without compromising its success. This explanation is not meant to imply that ethics ought to be compromised; quite the contrary, if a group has no room for any of these suggestions, then it may be that their measures of success and core values deserve reevaluation. Furthermore, it is an excellent exercise in self-reflection for a company to examine why it adopts certain safeguards and not others.

The first general step in this process is developing policies that articulate a banking company's position on the rights of donors. In making this decision, the interests of science and the company's fiduciary obligations must be weighed. At the very least, the policies should address the following points[35]:

1. Preventing individuals from being coerced into joining data banks
2. Preventing individuals untrained in genetics, both laypeople and health professionals, from making decisions based on genetic information, which they may misinterpret
3. Conserving medical resources by preventing genetics professionals and medical technologies from being used for inappropriate non-medical purposes
4. Conserving human resources by not disqualifying individuals from current activities (such as employment) because of a fear of future illness
5. Preventing discrimination based on genetic information
6. Preventing genetic reductionism and determinism, in which genetic explanation for health and behavior are the predominant factors in evaluating various aspects of human affairs

---

[35]Rothstein, Mark A. Genetic Secrets: A Policy Framework. In Rothstein, Mark (Ed.), *Genetic Secrets: Protecting Privacy and Confidentiality in the Genetic Era.* New Haven: Yale University Press, 1997. Annas, George J., Glantz, Leonard H., and Roche, Patricia. *The Genetic Privacy Act and Commentary.* Boston: Boston University School of Public Health, 1995.

7. Determine who may collect and analyze DNA
8. Determine who may inspect and obtain copies of records containing information derived from the genetic analysis of the DNA sample
9. Determine the purposes for which a DNA sample can be analyzed
10. Know what information can reasonably be expected to be derived from the genetic analysis
11. Address who may order the destruction of DNA samples. Delegate authority to another individual to order the destruction of the DNA sample after a donor's death
12. Determine what permissions are necessary for types of research or types of commercial activities

Although this may seem at first to be a difficult policy framework, addressing these points is not difficult because companies have already done so. Instituting the framework is relatively straightforward: standard operating procedures and policies already exist within a company; these suggestions need to be integrated into the legacy system in place. The true difficulty lies in whether management is comfortable with the answers that systematically address these points, and the reasons for discomfort if not. If these points were never addressed because the prevailing notion was to avoid a formal corporate response, then there promises to be trouble for the company in the future. There is a new dimension of forced transparency with every biotechnology company, and with that comes social expectations regarding corporate goals. Criticism will eventually find its way to any biotech firm that does not address its relevant bioethical issues; this point is one that was best exhibited in Chapter 2 on genetically modified foods, but the fallout from that industry should be a lesson to all companies dealing with advanced biotechnology. The public is sensitive to many of these issues, as are regulatory bodies like the FDA. It behooves any group to understand its position and develop policies, either privately or publicly, to clarify its core values and stance on these topics.

The next general step is one that will, like policy suggestions, recur throughout this text—the formation of an oversight board, like an institutional review board (IRB). Although IRBs are required by law for any type of research involving human subjects, their position is usually one of ap-

A Note on Gene Therapy

Biotechnology, indeed much of the Life Science industry, has an unwritten rule regarding diagnostic testing. Put simply, "Do not develop a test to diagnose a disease for which there is no treatment." Why? Because it is hard to sell the bad news that such a test might reveal if there is no way to treat it. With genetic information revealed by DNA data banking and efforts such as the SNP Consortium and the Human Genome Project, information is being accumulated that will produce genetic diagnostic tests. With all of the ethical issues facing this technology (privacy, insurability, employability, etc.), not having a therapy for a genetic disease predicted by such a test makes it extremely unattractive to consumers.

Fortunately, gene based drugs are being developed to lessen the ethical and health burdens associated with these tests. Although these treatments promise to alter the face of public health and drug discovery, they merely deal with the problem, not its root cause. To correct errors in disease-causing genes, a new technology has emerged, that of gene therapy. In its simplest form, gene therapy requires the identification of a faulty or disease-causing gene, and then the identification of the correct sequence. In effect, the normal sequence is inserted into an individual's genome to compensate for the diseased version. By modifying a retrovirus or adenovirus, which incorporates its own DNA into its host's genome to include a therapeutic sequence in lieu of its harmful sequence, the corrected version of a gene may be inserted. While the body may continue to use the disease-causing gene in its standard physiological activities, it will also transcribe the healthy gene. By doing so, the effects of the disease-causing gene (a malfunctioning protein) will hopefully be negligible in the presence of the inserted gene's product. For instance, a mutation in a gene that suppresses cell proliferation may lead to cancer; however, if the patient incorporates the wild-type sequence, then the new gene may regulate cell production, thus controlling tumor growth.

Although this method promises to someday "cure" genetic disease, many concerns surround this technology. At its very core, this, more than

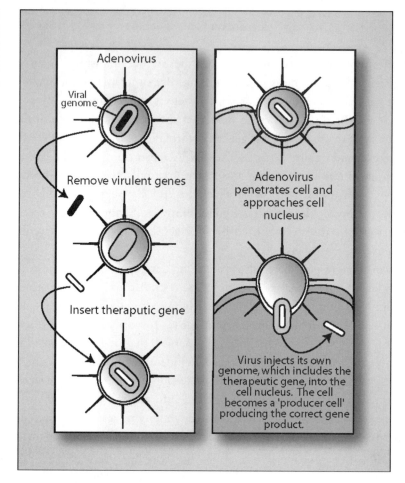

Figure 3-4   Gene therapy via adenoviral delivery.

any other genetic tool, facilitates the possibility of a new eugenics. Although warnings persist ranging from the history of Nazism to Huxley's *Brave New World*, the concern is difficult to dispel. Altering genes raises questions about our very definitions of normality. With the power to control phenotypes, how will we determine what is acceptable as therapy and unacceptable as enhancement? Is enhancement unacceptable? Granted, as a society, we do our best to improve the health outcomes and social status of our progeny, but is it okay to use biology as opposed to cultural mechanisms?

These are heavy questions, but more immediate concerns center on the safety of such technologies. In 1999, a very publicized incident occurred in which a gene therapy trial contributed to the death of an 18 year old, healthy volunteer. Jesse Gelsinger died after participating in a gene therapy trial at the University of Pennsylvania. Gelsinger was injected with an altered virus, which in its natural form may cause substantial liver damage; however, the engineered version should have been safe. To this day, the true cause of death is unknown, although most agree that the therapy was strongly implicated. However, criticisms abounded claiming that the death could have been avoided, and that the researchers were negligent. The most powerful criticism levied against the Institute for Human Gene Therapy (IHGT) at the University of Pennsylvania avers that the scientists withheld information from the patients and their families that would explain the level of risk associated with this study. Thus, consent was not informed.

Following a media blitz and an FDA investigation, the IHGT was ordered to temporarily suspend its human research programs, while obeying a list of other mandates. After reviewing the demands by the FDA, the IHGT decided to end all human trials and focus on animal experimentation. These are substantial setbacks to research, but they are not the only institutional challenges that this technology has brought to the research center. A lawsuit filed by Gelsinger's parents helped to maintain IHGT's presence in the media until a settlement was reached in late 2000. Although the details of damages paid are left undisclosed, the public relations damage is almost irreparable.

Following this event, a series of other high-profile tragedies followed. At St. Elizabeth's hospital in Boston, another gene therapy trial led to the death of a research participant. The ensuing media blitz called into question the study design, its consent procedures, and perhaps most damagingly, the commercial interests of the lead investigator. Similarly, in 2001, Johns Hopkins was forced to suspend all federally funded clinical trials, gene-based or not, after an asthma study led to the death of a previously healthy young woman. Sensitivity to these events may exist, but both industry and academia must truly learn from them, rather than simply ac-knowledge the tragedy because, to paraphrase George Santayana, being doomed to repeat them is unacceptable.

For those companies and institutions who recognize from this and oth-er examples the importance of responsible research conduct, the following suggestions will help in achieving this end. Most importantly, informed consent should accurately relay the level of risk involved, and it should take a dynamic form by updating participants on relevant safety data as it arises. Although this information can confound an experiment, it is im-portant to, at the very least, establish a threshold of clinical events that, if exceeded, would require notifying participants in the trial. To ensure such measures, increased oversight of trials with extreme attention paid to ad-verse events should inform the decision to report these incidents to the human subjects. Finally, care should be given to the types of applications. Experimentation should fit within carefully defined therapeutic interven-tions, rather than enhancement or germline engineering. If such experi-mentation is pursued, *substantial* input from ethicists and other con-cerned groups should be solicited.

proving research protocols, while addressing issues of potential harm to re-search subjects in a specific experiment. Because the nature of data banking involves maintaining biological material and data for many future applica-tions, IRBs should have a degree of authority over the corporate operations, stipulated by the Board of Directors. If that is too extreme or unnecessary for a company, the IRB should at least have a liaison to the Board whose communications enjoy a degree of urgency.

With respect to DNA data banking, the IRB should include particular types of members. The requisite doctors and researchers should be joined by ethicists, privacy experts, computer scientists and other information tech-nologists. Although there need not be one individual per skill set, all skill sets should be represented. Key to the success and objectivity of such a board is the inclusion of employees of the company and individuals who are not employed by the company. As such, the board can offer a pluralistic, objec-

Figure 3-5.

tive critique of experimentation and policy development. And although these suggestions may be too lax or too strict, they will offer a continuum of action in which conduct can balance corporate interests with social interests. An IRB that acts in this way proves itself valuable to the future of the company and its research, particularly when the future goals are unclear and may involve ethically challenging research. To best confront such issues, an IRB might be joined with an Ethics Advisory Board (EAB) to advise the company on specific issues and research goals. Examined in detail in Chapter 5, such a team may dramatically help a firm face the social challenges that go hand in hand with biotechnology.

To discuss the more specific ethical provisions that a company may adopt, an operations scheme based on drug development will be used. Because it is difficult to choose a single model for biotechnology product development, the familiar pharmaceutical scheme will be used instead (Figure 3-6). An important distinction is that this scheme may be used as a data banking company's actual operations, as in the case of DNA Sciences, which

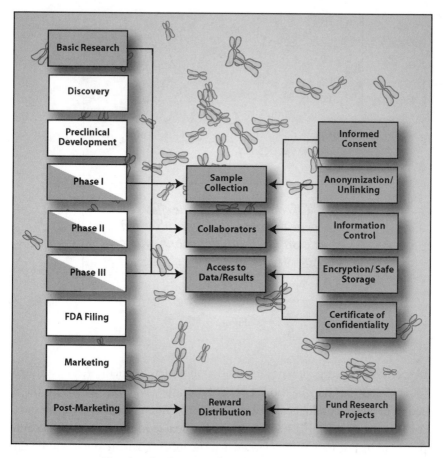

Figure 3-6   Strategies for ethical data banking.

performs its own research, or it may be used to articulate how a data banking company may fit with a corporate or academic partner, as with First Genetic Trust, which collaborates with Memorial Sloan-Kettering. In either case, the scheme is meant to illustrate the areas in which ethical tools can be applied, as well as which operational divisions are most susceptible to ethical concerns.

At the outset of any research involving human subjects, Federal regulations written over two decades ago require informed consent, which is intended to provide the research participant with the information necessary to make a rational choice regarding the research.[36] Although these regulations focus on conceptions of risks and benefits, recent trends have emphasized the process of informed consent as a bioethical tool that can be expanded beyond its legal role.[37] Since it is already a standard practice, many of the concerns previously discussed in this chapter have been incorporated into the informed consent process, which offers the perfect legacy procedure to adopt new, bioethical standards.

Representing Millennium Pharmaceuticals in the company's effort to work through these issues, Philip R. Reilly, Mark F. Boshar, and Steven H. Holtzman identify the areas where informed consent can be broadened toward more ethical concerns; they list the following points[38]:

1. A general description of the nature of the study
2. The identification and description of the research teams
3. The privacy guidelines for the study
4. Archiving
5. Distribution and other uses of the subject's DNA
6. Development of products for commercial gain
7. Other sensitive biological information
8. Consequences of findings of the study

To these points, the following should be added or offered for the sake of clarification:

9. Ownership
   a. Time limits
   b. Withdrawal procedures
   c. Conditions for destruction of samples
10. Proxy control stipulations
11. Permissions from group representatives

---

[36]Health and Human Services Regulations on Protection of Human Subjects. 45 Code of Federal Regulations, 46; 46 Federal Register, 8386 (1981).
[37]Reilly, Philip R., Boshar, Mark F., and Holtzman, Steven H. Ethical Issues in Genetic Research: Disclosure and Informed Consent. *Nature Genetics, 15,* 16–20, 1997.
[38]Ibid.

12. Recontact
13. Accessibility

Covering these points in the informed consent procedures allows the potential donor and the company to address both privacy issues and control issues. When these ideas are discussed in the following passages, the corresponding policy points should become evident. Essentially, the corporate choice to adopt or dismiss any of these positions should be included in its policies and/or operating procedures, thus reflecting the company's core values. They are presented as much as areas of inclusion under the informed consent procedures as they are as policy points.

The first two topics do not require much elaboration, as they are standard components of this process. Clearly, the third point is of tremendous importance. Explaining the privacy guidelines accomplishes two goals. First, the prospective donor understands to what extent his or her information will be protected, and, second, it assures the patient that care has been taken to address his or her interests. This latter detail will be one of the major issues facing this industry. Privacy concerns are not diminishing; indeed, with Internet sites like Amazon.com or DoubleClick constantly gathering information from visitors, privacy advocates will only shout louder when they sense potential transgressions against individual rights. Genetic technologies are often included in such arguments and the public is becoming increasingly aware of the growing presence of institutions that can invade their privacy. If they do not confront these important issues, businesses may find it harder to recruit donors, and industry and science will bear the consequences. It should also be noted that explaining these guidelines is much different than implementing them; this will be explained later in the chapter, in the discussion on accessibility.

With regard to archiving, point four, the donor should know where and what types of information are being stored. When searching for critical information regarding the whereabouts of material, it may be difficult to find an individual with the correct information. Although the archives may change location, and even ownership in some cases, an officer of the company should be chosen to maintain such information. Furthermore, contact information should be included in the consent procedure so a donor may contact the correct office in case of need.

The fifth point in this series is perhaps the most hotly debated topic regarding DNA data banking: the secondary use of biological material. Although material is usually collected for an express purpose, the scientific utility of a preserved sample may significantly increase in later, unrelated experimentation. For instance, *ApoE* is a gene that is involved in the metabolism of cholesterol, thus related to heart disease, but later studies found that it may also be an indicator for Alzheimer's Disease. While the donor knew that his or her material was being used for one type of research, consent for research on another disease may not have been given. With data banking, it is almost certain that secondary uses of stored material will ensue, thus the consent procedure must reflect the company's policy regarding this potentiality. Stanford University Professor of Law and Co-Director of Stanford's Program in Genomics, Ethics and Society, Henry T. Greely sees this as a significant issue, but not an insurmountable one. Greely offers an approach that allows institutions to confront this issue during the consent process:

> Permission for use of data or materials for unforeseen research is only valid if the person whose information or materials are to be used has been informed of the possibility of unforeseen future research uses of the information or material and the IRB review to be made of any subsequent unforeseen research uses. The person must have an opportunity to signify either the granting or denial of such permission. . . . The person must be informed whether his or her decision to agree to such unforeseen future research uses will affect his or her participation in any such specific research or in any program activity. The grant of such permission must not be influenced by any undue inducements to the subject to participate.[39]

The use of material for commercial gain, point six, is fairly self-evident. Requiring the disclosure that the tissue or data may result in commercial gain is redundant in the corporate context; however, it is still necessary to in-

---

[39]Greely, Henry T. Breaking the Stalemate: A Prospective Regulatory Framework for Unforeseen Research Uses of Human Tissue Samples and Health Information. *Wake Forest Law Review, 34,* 754. 1999.

clude this part of the consent form. The donor should be informed as to who stands to gain from these experiments, which may extend beyond the company collecting material. Obviously, current collaborators come to mind, but provision should be made for the potential of future collaborators, who may not be known at the time of collection. A statement that explains that partners are actively pursued to share in the discovery and product development cycle should suffice.

In the case of the seventh point, other sensitive biological information, the emphasis lies in how to handle information that impacts more than just the donor. In more than one instance, information has been discovered that may reveal nonpaternity or that an individual was adopted, as well as other information that relates to the donor's kin. Although this is rare, and it is more pertinent in genetic testing, where clinical genetic analysis results are always given to the patient, it is important to discuss it in the context of data banking as well. A company's policy regarding access to information by a donor's family should be disclosed in the consent procedure.[40] Although each company will have its own policy, it should be examined against the core values of the company, which in turn ought to be developed with bioethics in mind. According to the Genetic Privacy Act (GPA), a proposed legislative framework for these very issues, the sound solution places the answer within the bond of family, leaving the decision to the individual's own sense of familial responsibility. The GPA opts for protecting privacy by prohibiting an investigator from disclosing information to others:

> Philosophically, the genetic information is and should remain private . . . people . . . have a right not to know as well as be the only one who knows. Moreover, carrying a particular gene does not put anyone else at risk. Family members have a moral obligation to other family members, and should be encouraged to share genetic information that may have an impact on other family members. . . . As a practical matter, defining the scope of such a duty would be virtually impossible.[41]

---

[40]Macklin, R. Privacy Control of Genetic Information. In *Gene Mapping: Using Law and Ethics as a Guide.* New York: Oxford University Press, 1992.

[41]Annas, George. *Some Choice.* New York: Oxford University Press, p. 110, 1998.

However, the company should reflect and determine whether there are any conditions upon which they believe disclosure to family members would serve the greater good. Initial efforts should focus on counseling the individual to inform the family, but other circumstances might require notification at any cost. Although this is a precarious position to take, given that it may violate privacy and even a contract, the firm must decide whether knowing that an individual's family members may have or do have a treatable disease (for instance, if the disease only requires one allele, and the subject carries two copies, then his or her brothers or sisters have at least a 75% chance of having the disease by simple Mendelian genetics) requires notifying the family. Perhaps the corporate policy/consent form should account for this and similar cases; for instance, if the analysis reveals a disease that has a known, effective therapy, then it is the obligation of the investigator to participate in a means of notification and recommend treatment to the person's physician. The policy can be attenuated so it only applies to diseases of grave consequences; thus, the severity of the disease will be another consideration in the decision to notify people who have not participated in the trial.

The eighth point, consequences of findings of the study, addresses the most controversial aspects of data banking. Information may be revealed that can compromise the insurability or employability of the donor. Since this information ranges in its applicability from complete predictive value to suspected risk, the value of this information to these exclusionary decisions may be unsubstantial. Regardless of the company's stance on the actuarial utility of genetic information, that there is a risk associated with genetic analysis should be disclosed. In the interest of fairness to the donor, the potential to be penalized for contributing to science ought to be presented.

Ownership, the ninth point, represents a great tension in the balance of ethics and corporate interests. Is the DNA data bank a true bank, where a sample is deposited and held in trust until the donor decides to withdraw it, or is it only a bank in name, where the company owns the material, while affording some level of control to the donor? In either case, the corporate policy on this topic should be well thought out, as the presentation of itself as one type of organization, while actually being another is a potential public relations nightmare. Furthermore, the donor deserves to know his or her rights with regard to the material. Another event similar to the case of John

Moore is inexcusable in today's business climate, and public interest groups will make sure that the issues are handled sensitively.

Under the ninth point of the consent process, a corporation ought to address time limits on the storage of material and withdrawal procedures (destruction procedures). Prior to donation, the donor should be informed of the standard policies for storage of material, and whether intervals exist when the donor will be contacted, or when the donor should contact the company to decide on the "withdrawal" of material. These should be concrete numbers, and the reasoning behind recontact ought to be reviewed by the IRB prior doing so. In the case of extensive storage, "An initial request for an indefinite lifetime for the research use of the information or materials may be approved by an IRB only upon showing that it is appropriate under the circumstances. . . . Any extension of the lifetime for research use of the information or materials collected beyond the period set out in the initial consent or permission may be approved by an IRB only upon a showing of some extraordinary particular need for or benefit from such extended use."[42]

In terms of withdrawal procedures, provisions similar to archiving should be made. If the donor is afforded the freedom of sample control, and thus the ability to remove material *and* data, or material *or* data (according to some FDA regulations, it may be impossible to remove data for some trials), then the contact information for the corporate representative responsible for this activity should be made available. Furthermore, this individual or the corresponding office should be easily identifiable and accessible to donors.

With respect to the control of samples, the tenth point provides for the inability of the donor to make decisions that the consent process states will be necessary in the future. In the unfortunate event of the donor's passing or if he or she is rendered unable in some way to make decisions on his or her behalf, then someone should be identified as a proxy during the consent process. This person is given the same level of control over the material as the original donor after the donor forfeits that ability due to whatever extenuating conditions force this end.

In terms of the penultimate point, permissions from group representatives, the concern revolves around information or conclusions that may be

---

[42]Greely, Henry T. *Breaking the Stalemate*, p. 755, 1999.

taken from the research and implicate more than the individual's immediate relations. Investigating a single disease may affect an entire ethnic group if, for instance, there is a prevalence of the disease among them. An example might be the aforementioned sickle cell anemia, which has a high incidence in individuals of recent African origin. As mentioned above, the revelation of this disease risk in populations of African-Americans resulted in exclusionary practices in the United States Air Force. The genetic predisposition was valued negatively immediately, while a single copy of a gene for sickle cell has been beneficial historically. In what is called a balanced polymorphism, the gene responsible for sickle cell, *hgB,* like most genes, exists in two copies in every human. The mutation responsible for sickle cell confers resistance to malaria and no deleterious effects when it is accompanied by the normal sequence.

The potential misuse of this information can be substantial. As such, it ought to be disclosed to any potential donors. The difficulty alluded to earlier involves consent from more than just the individual donor; representatives from these populations should also be consulted. In a society as diverse as the United States, it is difficult enough to pinpoint an actual group, ethnic or otherwise; finding adequate representatives is an even greater challenge. Too many different world views complicate this matter, but efforts should be made to engage some level of discourse with institutions that claim to be representative of the demographic group in question.

Recontact moves the discussion beyond the immediate consent process. In terms of secondary uses of material, the clause refers to future, unknown use of the material. Because the future is uncertain, a consent form cannot spell out the exact conditions, experiments, and collaborators to which the sample may be subjected. To be as clear as possible, and equally honest, it is necessary to state that the material *might* be used in research that is unknown at the time of donation. This disclosure does not and cannot accomplish the impossible task (few consent processes have that psychic ability) of identifying future risks; it merely states that they may exist. Because of this ambiguity, there ought to be a provision in the consent process whereby the donor and the investigator agree on the circumstances and methods of recontact in the event of the secondary use of material. This position is clearly linked to point nine, or ownership, where the temptation is to award the property right to the company, thus avoiding any scenario where recontact is necessary. This alternative should not be tacit, but rather offered as a choice to the donor, irrespective of the ownership issue. During the consent process, the donor can be given the choice as to whether he or she wants to be contacted with regard to future use of the material.

There are acceptable, ethical conditions upon which recontact can be avoided. To communicate with a donor, certain information is necessary, for instance name, address, and other identifiers. Furthermore, this information must be linked to the samples. If the samples were not linked to this information, however, then not only is recontact unnecessary because it is impossible, but many of the ethical concerns regarding privacy rights are obviated. During the consent process, the potential donor may be given the option to participate on the condition that his or her biological material will not be linked to any information that is identifiable. Indeed, the protections offered by the use of unlinked material has led to "conformity between nations about raising a distinction between anonymous research where the identity of the subject is not revealed and where the research is conducted on the basis of personal data. Only in the case of the latter is there a requirement of informed consent."[43] Thus, acceptable, ethical global business practices can be derived from total unlinking of samples from personal identifiers.

With respect to labeling samples and recontact, NBAC has developed useful guidelines on how to include these measures in the informed consent process. They suggest the following options be presented to the potential donor:

1. Permitting only unidentified or unlinked use of their biological materials in research
2. Permitting coded or identified use of their biological materials for one particular study only, with no further contact permitted to ask for permission to do future studies

---

[43]Jonatansson, H. Iceland's Health Sector Database: A Significant Head Start in the Search for the Biological Grail or an Irreversible Error? *American Journal of Law and Medicine, 26* (1), 54, 2000.

3. Permitting coded or identified use of their biological materials for one particular study only, with further contact permitted to ask for permission to do future studies
4. Permitting coded or identified use of their biological materials for any study relating to the condition for which the sample was originally collected, with further contact allowed to seek permission for other types of studies.

Unlinked information and coded (anonymized) information are very different forms of storage. In the case of the former, there is no way to link any information to a sample, or the individual who donated the sample. In the case of coding, samples are anonymous to most people who come in contact with them, but someone has a master key that links data, the material and the donor. This area is somewhat murky in terms of maintaining informational privacy, with complications due, in part, to the collaborative nature of science. Although it may appear that the only institution that has access to information is the one collecting the material, it may not be clear to the donor if collaborators will have access to the information. During the consent process, the terms of joint research have to be disclosed; specifically, what, if any, types of information are accessible by business and scientific partners. Furthermore, it is important to do a degree of due diligence and audit collaborators regarding their practices to protect privacy in these cases. If none are in place, it will be necessary for the data banking company to limit the flow of information to make sure that the rights of the donor are protected to *at least the extent* that the company had promised the individual in the initial consent document.

Accessibility is a topic that must be reflected in the consent process, but it should be established in a policy framework. Before deciding on how a collaborator should respect a donor's privacy, the data bank must establish its own operating procedures. First, the company must decide under what circumstances and who may have access to identifiable information. The decision should be made with the aid of an IRB or EAB, which can help decide what conditions make accessibility absolutely necessary. It is not enough to state these policies; penalties for violations must also be established and enforced. To aid in this process of protection, coding systems should be employed, as well as password security on databases and computers. Furthermore, information should be stored in physically and communicatively separated areas—different, locked file cabinets and different computer servers.

Many of these practices can be avoided if the information is unlinked; however, the computer age makes it difficult to proceed with anything without creating a digital record of its occurrence. To ensure such unlinking measures, however, one-way encryption methods should be adopted. These processes are often called strong encryption. In this case, information is input into a computer and parameters are set as to what type of data is to be stored with the sample, and what is not. The program then strips the sample and data of its identifying information, but rather than coding the identifying information and placing it in a separate file, it is discarded. Thus, it is impossible to link the data to an individual.

There is also the matter of allowing access to the donor. In many cases, it only seems fair that the donor should know as much about his or her health as anyone else; thus, if a company has genetic knowledge about someone, it does not follow that the individual should remain ignorant of the matter. There are reasons for keeping information from the donor; the most powerful is to keep the donor from doing harm to him or herself. If the research participant is unaware of information that might exclude employment or insurance, then he or she cannot compromise either. Still, the case is quite strong that the individual should have the ability to access, purge, and update information derived from experimentation on his or her biological materials. Of course, in cases where a company offers banking services for storage or for future health use of the donor, this will be expected. In other cases, new, accurate information can only aid in experimentation, but the ethical cost of informing the patient must be weighed against the type of research performed. For instance, if the results of conducting tests lead towards the diagnosis of an incurable disease, the utility of the information comes into question.

All of these privacy safeguards are meant to protect the dignity of the donor by not violating his or her trust, but there are more tangible concerns that they address. No one wants to donate material and have the results of this process be used against him or her, either in a discriminatory or a legal capacity. This information can be released in paternity cases or criminal investigations upon subpoena. As an act of faith, it would go a long way for a company to ensure that such intrusions can be prevented. To do so, a U.S. company can apply to the Department of Health and Human Services for a Certificate of Confidentiality. These certificates allow for the following protections:

> The Secretary may authorize persons engaged in biomedical, behavioral, clinical, or other research (including research on mental health, including research on the use and effect of alcohol and other psychoactive drugs) to protect the privacy of individuals who are the subject of such research by withholding from all persons not connected with the conduct of such research the names or other identifying characteristics of such individuals. Persons so authorized to protect the privacy of such individuals may not be compelled in any Federal, State or local civil, criminal, administrative, legislative, or other proceedings to identify such individuals.[44]

Certificates of confidentiality have explicit provisions that protect the release of genetic information, even by subpoena. This is a one-way document; it prevents the forced disclosure, not the voluntary disclosure.

The final recommendation in this process is perhaps the most extreme test of appreciation for donors. As the *Moore v. UC Regents* case indicates, there is a great deal of controversy over the concept of ownership and proprietary privacy in this scheme, and to whom profits from DNA databank research should be distributed. Although the company doing the research invests a substantial amount of time and money, many of the advances are predicated upon the financially unrewarded efforts of individuals willing to relinquish their genetic material in the name of science. As mentioned earlier, there is the incentive of knowing that research is being done on genetic conditions afflicting the donor and/or his or her family, although the treatments may not materialize in the donor's lifetime, if at all. Even if a company were to involve itself in profit sharing with tissue donors, there is

---

[44]Public Health Service Act §301(d), 42 U.S.C. §241(d). *http://ohrp.osophs.dhhs.gov/human-subjects/guidance/certconpriv.htm.*

difficulty in determining who deserves what—should anyone with a genetic disease get more than those with a history of disease but no genetic component? The negative controls are as important as any other part of the cohort, but should they receive remuneration? To still give back without having to contend with solving irresolvable problems like these, the best method is to honor the intentions of the donors. In most cases, these people have donated material to either help themselves through aiding efforts for therapy, or help science relieve the plight of a diseased population. In either case, to honor the spirit of tissue donation, the companies who profit from this research can give back by donating money to researchers and social groups dedicated to relieving the suffering of those people who have the disease in question. For instance, if DNA is collected from diabetics, and the resulting research leads to a profitable drug, then some percentage of profit can be dedicated to internal and independent research efforts and/or the American Diabetes Association. Such initiatives highlight the social responsibility of companies, and reinforce their dedication to aiding society through biotechnology, as in the case of Warner Lambert, which instituted this very profit redistribution policy for one of its off-patent drugs.[45]

Although these recommendations offer a framework for dealing with the issues that surround DNA data banking, there are more wide-reaching concerns that have not been addressed. Many of the initiatives mentioned thus far are aimed at First World problems, which are echoed in developing countries, but in cases of the latter, there may be different emphases or more profound effects of negligence. Recently, the current state of bioethical conduct in the world was the topic of a six-part series in the *Washington Post*.[46] These articles condemn many of the practices of First World drug and biotechnology companies using inhabitants of developing and underprivileged areas in foreign countries for drug trials. Although the findings are abhorrent, there are investigations underway to confirm their validity (the reports seem very antiindustry and do not devote very much text to industry's

[45]Cook-Deegan, Robert Mullan. Confidentiality, Collective Resources, and Commercial Genomics. In *Genetic Secrets,* Rothstein, Mark (Ed.), New Haven, CT: Yale University Press, 1997.
[46]The Body Hunters. *Washington Post*. December 16–22. 2000.

response). Nevertheless, the series points to a very important problem: ethical safeguards can be easily overlooked in poorly regulated areas, particularly if they are far away from the company. The article makes the striking point that "these [bioethical] principles have been breaking down as drug companies enroll thousands of test subjects at a time in Eastern Europe, the former Soviet Union, Africa, China, Latin America and elsewhere."[47] Group consent has been discussed above, but the case of foreign cultures deserves particular attention. The informed consent process only works when there is an emphasis on "informed." The formal nature of this device should not ever, foreign or domestically, reach the point where it is seen as a protection for the investigator rather than the subject. Very disturbingly, "a recent survey by two Johns Hopkins University researchers of more than 500 U.S. and foreign investigators found that most thought consent procedures in experiments were mainly used to shield researchers from lawsuits, rather than to protect patients."[48]

In one of the *Post*'s stories, research in China is cited as a case in point. Millennium Pharmaceuticals, a company firmly committed to ethical business, jointly funded a research project with Harvard University to obtain and investigate an isolated, genetically homogenous population in Anhui province. Although Millennium and Harvard have strict ethical guidelines, the question still arises: "Are some populations too vulnerable for all but the most essential medical research? . . . 'We were very mindful of having the same standards applied to them as in the U.S.,' said Harvard Provost Harvey V. Fineberg. 'Every effort was made to assure that was the case.'"[49] However, recruiting donors from this population was left to local officials involved in the project. Informed consent took the form of handing out written documents explaining rights and the conditions of research, a practice that works well in the United States. Despite the initiative, little was done to explain the actual content of the pamphlets to the predominantly illiterate population, and many of the potential donors did not care because they were con-

---

[47]LaFraniere, Sharon, Flahety, Mary Pat, and Stephens, Joe. The Body Hunters: Part 3, The Dilemma: Submit or Suffer. *The Washington Post,* December 19, 2000, A01.
[48]Ibid.
[49]Pomfret, John and Nelson, Deborah. The Body Hunters: Part 4, In Rural China, a Genetic Mother Lode. *The Washington Post,* December 20, 2000, A01.

vinced that they had much to gain from participation. Healthcare in the province is minimal, and the doctors who advertised the trial did so using discounted health care as an incentive. "Those who participated were to get a free exam, test results, follow-up care and a 'health card' for a discount health care program."[50] Alleging that these inducements were never honored, the *Washington Post* highlights a very important concern: bioethics must become a way of thinking, not just a set of regulations. In this case, researchers relied on local investigators with an interest in the research to handle the most sensitive aspect of the experimentation—human subject recruitment—which resulted in possible violations of human rights. Furthermore, the cohort's lack of comprehension and the alleged nondelivery of promises invokes one of the most powerful critiques against researching in developing countries: are corporations and other researchers "stealing" DNA? Born out of discussions of intellectual property rights, the previous example, if it is substantiated, confirms the invasion of privacy that many individuals fear in cases where underprivileged groups are exploited in the name of research. It is critical to understand that the only condition of human rights is that they extend to every human.

The lesson here is for investigators to make concerted, involved efforts in the bioethical systems that relate to a study, and not to rely on the conscience of others, particularly in unfamiliar cultures. The role of an IRB is critical in these types of experiments, and cultural liaisons without any sort of interest in the research ought to be actively involved in ensuring the ethical treatment of research participants. It should be noted that all of these steps were taken by Millennium Pharmaceuticals as soon as these transgressions whether founded or not, were brought to their attention. Because it will take more than standard forms to adequately inform participants, counselors and scientists alike should be involved in presenting information and answering questions. Furthermore, psychological experts can be invited to ensure that individuals are fit to make a decision as important as relinquishing a "possession" as valuable as DNA.

There is a directly proportional relationship between the scientific and the cultural value of DNA. As genetic research reveals more information

---

[50]Ibid.

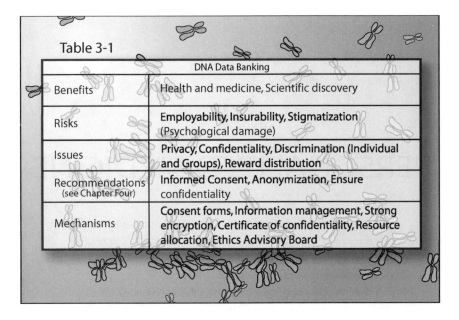

Table 3-1

DNA Data Banking	
Benefits	Health and medicine, Scientific discovery
Risks	Employability, Insurability, Stigmatization (Psychological damage)
Issues	Privacy, Confidentiality, Discrimination (Individual and Groups), Reward distribution
Recommendations (see Chapter Four)	Informed Consent, Anonymization, Ensure confidentiality
Mechanisms	Consent forms, Information management, Strong encryption, Certificate of confidentiality, Resource allocation, Ethics Advisory Board

about the human genome, there will be more interest in gathering samples and people will have greater interest in their own genetic blueprint. As knowledge expands the realm of genetic applications, so too will the ethical questions expand. While the benefits to science and humanity may be substantial, it would do corporations well to appreciate that an individual's genes are as precious to its source as they are to biotechnology.

4

PERSONALIZED MEDICINE

Human knowledge and human power meet in one; for where the
cause is not known the effect cannot be produced. *Nature to be
commanded must be obeyed;* and that which in contemplation is as
the cause is in operation as the rule.

—Sir Francis Bacon, *Novum Organum*

I. Executive Summary

Medicine has always advanced with the aid of science, but it is practiced as an
art. There are no equations or axioms that immediately predict how to best
treat a sick patient. It is a game of odds, where past experience informs the
present, and while the same symptoms in one person require aspirin to relieve
a condition, another individual may need a lipid-lowering drug. By trial and
error, our one size fits all mode of therapy eventually and hopefully will cure
the diseased. However, the art of medicine has accomplished many impressive
feats, ranging from putting a patient's cancerous condition into remission to
creating vaccines to protect against some of the most virulent diseases. Still,
these methods only work in a subset of a diseased population, whereas the
true goal is to target the largest possible number of patients, preferably all of
them. Human variation remains the challenge to medicine, and until science
can harness rather than compete against it, care can never be delivered at an
individualized level. Recently, efforts related to decoding the human genome
have revealed one area in which variation can be used for the betterment of
human health. Personalized medicine will be the realization of using an individual's genetic information to choose the best course of therapy for him or
her.

The Science

Personalized medicine relies on understanding the genetic variations that influence an individual's response to therapy. Using technologies known as pharmacogenomics and pharmacogenetics, information based on knowledge of an individual's genes can be used to identify the conditions by which drugs will be most effective.

The Benefits

These methods will not only increase the speed and efficacy by which drugs are developed, they will also decrease the costs associated with pharmaceutical research. The resulting products will be targeted at individuals to increase response rates to therapy, while decreasing side effects. Time will no longer be wasted on trying a series of drugs for a condition until one works; rather, the appropriate therapy will be prescribed based on the genetic and biochemical individuality of the patient. This efficacy of care will translate into cost savings to patients and healthcare providers, while cost-effectively improving overall public health.

The Issues

Genetic information of any sort is value laden. Although the positive aspects are often overlooked, the negative ones are usually amplified. For this reason, the questions concerning psychological damage, gene patenting, and global healthcare inequities are examined in this chapter. The reader would do well to reexamine the preceding chapter, since many issues dealing with data banking are intimately related to this discussion.

The Industry

Although this technology is very promising, it also has immediate economic repercussions on large pharmaceutical companies. Currently, these companies have many therapies on the market, and personalized health-care may reveal that many people on these drugs ought not be taking them. Cutting into the market would compromise the economic imperative of these firms, and despite their overall acceptance of the technology, bottom lines may dictate that its use be postponed. However, the biotech sector is poised to promote the use of the technology, regardless of the reluctance of the pharmaceutical industry. By doing so, a company like the one profiled in this chapter, Interleukin Genetics, assumes the role of patient advocate because their success is directly linked to the success of the treatment.

Recommendations

To address the ethical issues associated with this technology, many of the points brought up in Chapter 3 should be revisited. Beyond those steps, efforts can also be directed to employ genetic counselors to educate and advise patients or clinical trial participants. Education should expand beyond immediate stakeholders to all those affected by the technology as well. Such efforts, critical components of marketing, need to be as objective as they are informative.

Recently, television commercials have been contributing to public health and pharmaceutical profits in a new way. Prescription drugs are being advertised in the media to inform the viewer/patient/customer that common illnesses can be treated with new therapeutics. Allergy sufferers need only ask their doctor for the advertised treatment by name, and in all likelihood, the patient's physician will prescribe it. While extolling the virtues of these pharmaceuticals, these commercials are required by the FDA to offer something new: warnings of side effects. Because these disclosures are novel to drug advertising, they may receive more attention than the drug itself. However, pharmaceutical companies still use direct-to-consumer advertising because they realize that most doctors will prescribe name brands if a patient requests a particular drug.[1]

Truth in advertising represents just one of many factors that are transforming the role of medicine in society. Indeed, this example scratches the surface of one of the most promising revolutions in patient care, although it may not seem completely obvious at first glance. Below the surface lies the revolutionary concept of personalized medicine—delivering individualized care based on a patient's genetic profile. Consider a standard advertisement for an allergy medicine in which a child dances through a patch of flowers that would otherwise render her traumatized by swollen eyes, irritated sinuses, and possibly even hives. At the close of the ad, against the backdrop of daisies, a confident voice explains that side effects may include unusual drowsiness, higher sensitivity to bruises, or yellowing of skin to name but a few. "Check with your doctor," the voice concludes, confidently assuring the viewer that it is worth the effort to at least try the drug. And many people do, but why? Two reasons stand out in answer to this question. The first centers on probability; one may ask, "what are the chances that I'll respond to the drug without side effects?" Conventional wisdom holds that not everyone experiences adverse effects, so why not give it a shot? Beyond this explanation is the growing number of informed consumers, more appropriately, empowered consumers. Information abounds in the age that now shares the name of this commodity, and in our Information Age patients can take it upon themselves to choose their own therapies and research their own conditions. The idea that an individual can take control of his or her own medical needs is reassuring, and easy access to experimental information as a result of the Freedom of Information Act, which enables any U.S. citizen to access clinical trials data, reinforces the educational foundation necessary to control one's clinical fate.[2] Indeed, the role of explaining and interpreting disease rests less and less with physicians as patients assume greater responsibility for their own health. After all, who could know someone's body better than the individual herself? Bioethicists who recognize this fact are now pon-

---

[1]Hodnett J. Targeting Consumers. *Medical Marketing & Media,* November 1995, 91–95.
[2]*www.fda.gov/foi/foia2.htm*

dering whether some diseases are treated better by the patient than by the doctor.[3]

Medicine has always been personal, but it is about to become even more so. Side effects may only occur in certain individuals, and to relieve the symptoms of a disease, many are willing to make the bet that a drug will deliver substantially more positive effects than negative ones. However, new technologies are being developed to remove the guesswork surrounding drug response, honing therapies to particular individuals. Hay fever will no longer be replaced by nausea when a simple cheek swab to collect DNA for a low-risk test reveals that a name brand will relieve a patient's symptoms without causing an adverse reaction, whereas a generic trades symptoms for side effects. Reducing side effects is just one of the many ways that personalized medicine will improve therapy. Using new technologies, known as pharmacogenomics and pharmacogenetics, to develop drugs based in part on genetic traits of a portion of the population not only reveals more precise therapeutic strategies, but also correlates to variability in the genes used for their discovery. If a drug is developed specifically to block the activity of the product of a particular gene, as in the case of Novartis' new drug Gleevec, then alterations in that gene may control the pharmaceutical's efficacy.

Even before studies are performed to link genetic variations to drug response, linking drug development with knowledge of genetics substantially aids public health and industry. Gleevec is targeted at a particular gene critical to the pathology of chronic mylogenous leukemia (CML). A variation in the gene known as *abl* causes the uncontrolled proliferation of white blood cells, leading to CML. Knowing the specific gene involved with this disease allowed researchers to develop a treatment tailored to the genetics of the disease. By attacking the protein transcribed by this gene, Gleevec's therapeutic precision is substantially greater than any current treatment strategies, so long as the illness is caught early enough. As Stanley Nelson of UCLA explains, "Tumors that look exactly the same can respond dramatically differently to chemotherapy and radiation. . . . But we've had to treat them as if

---

[3]Lantos, John D. *Do We Still Need Doctors? A Physician's Account of Practicing Medicine Today.* New York: Routledge, 1999.

they were all the same." He says that with a full genetic catalog of tumors, "we ought to be able to attack them in a far more effective way."[4]

While Gleevec exemplifies the promise of personalized medicine, it also displays the technology's youth. Targeting genes is effective, but there still are other variations that may affect disease pathology and drug response. Many diseases are polygenic; that is, multiple genes affect them. One or multiple genes may be causal, while others may amplify or dampen the disease. In the case of Gleevec, the drug does not work as well as everyone hoped, but it fits a known pattern. Some people respond tremendously well, while others do not. Furthermore, individuals may respond well to treatment at the outset, but over time, the efficacy of the drug may diminish. If anything, the variable response to genetically derived therapies further shows the importance of individualized care.

Until recently, the attention lavished on the Human Genome Project has centered on the discovery of genes that when mutated cause genetic disease. Those with greater foresight also saw this accomplishment as a means to streamline the drug development process by identifying therapeutic targets based on a thorough understanding of the human genome. However, those Promethean thinkers who believed in an even greater payoff saw that unlocking the human genome offered a revolutionary means of helping patients; the Human Genome Project is the first step toward understanding not only how to use individual genetic profiles to predict more than just disease, but also the drug cocktails that a specific patient would need to take to optimally treat disease. Doctors have always known that each patient is unique, and what is more unique than an individual's DNA? What better way to personalize medicine than to leverage the information derived from a patient's genome?

Although it may seem logical that a drug developed based on genes will reach the market hand-in-hand with the ability to carry out the same genetic profiling that led to successful clinical trials, there are many drugs that have made it to market before the technology that allows correlating genetics to differential drug response. Because the therapies were released prior to

---

[4]Begley, Sharon. Made to Order Medicine. *Newsweek,* http://www.msnbc.com/news/588559.asp.

understanding the nature of pharmacogenomics and pharmacogenetics, it may take some time before such analyses precede their prescriptions.

In addition to such technical challenges to the rapid introduction of personalized medicine, pharmaceutical companies' resistance to this new mode of treatment also impedes its progress. In many cases, drugs amenable to pharmacogenetic targeting of responsive individuals have already established a strong market share. Introducing genetic information may limit the market appeal of a particular therapy if the percentage of genetically determined responders represents a smaller population than the drug's established market penetration. With little economic incentive to release their data supporting personalized medicine, "Big Pharma," pharmaceutical companies with market capitalizations in the many billions of dollars, will also be reluctant to do so because the information may be too preliminary in their eyes.

To further its scientific and economic goals, Big Pharma is pursuing pharmacogenetics, but instead of focusing on drug response profiles, many firms focus on genetic predisposition to side effects. Because patients who suffer side effects represent both health risks and liability to a company, reducing these events translates into a balance between the profit motive and public health. For many critics of the technology, pursuing this research agenda is a forced concession following a recent public health and public relations debacle for one of the major pharmaceutical companies. In late 1998, Smith Kline Beecham (now GlaxoSmithKline) released LYMErix, a vaccine for lyme disease. However, within six months of its release, over 500 adverse events were reported to the FDA, 45 of which were considered serious, and three resulted in death. A class action lawsuit, as well as other suits, followed, and "a major allegation in the suits is that individuals of a certain genetic type, known as HLA-DR4+, are susceptible to developing an incurable form of arthritis if they are exposed to the protein that makes the vaccine work."[5] However, GlaxoSmithKline claims that although many of the patients in the clinical trial cohort carried the HLA-DR4+ genotype, none experienced the autoimmune side effects.[6] Although the actual data are yet to be released,

[5]Warner, Susan. Patients Sue Over Effects of Vaccine. *The Philadelphia Inquirer,* June 13, 2000.
[6]Noble, Holcomb B. Three Suits Say Lyme Vaccine Caused Severe Arthritis. *The New York Times,* June 13, 2000.

and it is unknown if patients who participated in the clinical trials were actually genotyped, the suits are still pending. Either way, personalized medicine is gaining a level of great prominence in the life science industry, and it is hoped that its fame remains in this sector, not the courts.

Interestingly, pharmaceutical companies are not the only players in this field. New companies are emerging and focusing their research on pharmacogenetics or pharmacogenomics. These biotechnology firms are assuming a risky position as they challenge the ethics of the pharmaceutical industry while confronting the practices of their own research. Comforted by the knowledge that their corporate success is tied to patient advocacy, companies dedicated to personalized medicine face a host of internal and external challenges that, if overcome, may revolutionize medicine.

II. Science

The technology behind personalized medicine falls into two main categories: pharmacogenetics and pharmacogenomics. These approaches have the shared goal of personalized medicine, despite their different strategies, and although they overlap in some approaches, their differences affect how they each contribute to healthcare. On one hand, these technologies aid drug discovery by identifying novel physiological targets for pharmaceuticals; on the other, they provide knowledge that dramatically improves the drug development process from start to finish. Pharmacogenomics uses DNA-based data derived from large patient populations to achieve this goal, whereas pharmacogenetics relies on specific genetic aberrations instead. The former is primarily useful in drug discovery and development, but the latter approach can be applied beyond discovery and utilized in a clinical setting. Because specific genetic variants are linked to the drug discovery process, they can also be used in a genetic test to provide the knowledge necessary for a physician to assign an appropriate course of therapy.

Although they are very similar, sharing the identical goal of personalized medicine, they do employ subtly different strategies of achieving their ends. However, after explaining their distinctions in this section of the text, to avoid further confusion they will be treated synonymously because outside of a research context their similarities are strong enough to take their differences for granted. These technologies use many of the same tools, and rely on much of the same information; however, the idea that each approaches drug discovery, development, and delivery differently should not be forgotten. In both cases, the techniques rely on genes to identify potential drug targets. After isolating targets based on genes, pharmaceuticals are developed to react with these specific targets, which is followed by extensive testing for efficacy and safety. Environment and biology, that is, drugs and genetics, are investigated interdependently, resulting in treatments in which drug response is correlated to genetics.

Allen D. Roses, Director of GlaxoSmithKline's Pharmacogenetics Division and former Chair of Duke University's Genetics Division, best describes the difference between these strategies for drug discovery and development: "Discovery [pharmaco]genetics uses human disease populations to identify disease-related susceptibility genes. Discovery [pharmaco]genomics uses the increasing number of databases of DNA sequence information to identify genes and families of genes for tractable or screenable targets that are not known to be genetically related to disease."[7] In the "genetic" version, investigators focus on genes that are already implicated in disease. If a variant of this gene, known as a single nucleotide polymorphism (SNP), translates into a functional difference between this version and the "normal" version, then the effects of this change are investigated to identify clinical significance. Unfortunately, the process is not so simple as creating therapies that interfere with the disease-related gene's product; that is, when the gene creates a protein, that protein may not be an acceptable target, regardless of its role in disease. Drug targets are limited to a few types of proteins because of the deleterious effects that certain drugs can have on the body. Targeting a receptor on a cell or a part of a cell is better than targeting a type of cell because the former is more specific, whereas focusing on the latter may attack more than just disease. Also, extracellular protein targets, for instance those on the outer membrane, rather than proteins that are inside the cell, make better targets because they are not inside a protective membrane. In cases

---

[7]Roses, Allen D. Pharmacogenetics and the Practice of Medicine. Insight Review Articles. *Nature, 405,* 858, June 15, 2000.

where the proteins of the susceptibility gene or genes are not good candidates for therapy, the expressed protein is analyzed in a broader context. Because it is most likely part of a pathway, the protein may turn genes on or off, or modify other proteins. If interrupting all or part of this pathway leads to disease, then the metabolic target can be narrowed. Pinpointing where the error occurs may reveal another means of correcting the problem, either by introducing a missing protein or blocking the expression of another. Thus, studying the genetics of a disease can reveal the best drug candidates, and since the particular genetics behind the disease informed the discovery process, that information can be used to correlate the drug's efficacy to a particular genotype. (See Figure 4-1.)

The "genomic" strategy does not have the advantage of designing treatments based on disease genes; rather, it takes the reverse path. Whereas pharmacogenetics begins with a genetic variation and then proceeds towards treatment, pharmacogenomics starts with a potential treatment and then correlates it to genes. During a drug study, a battery of tests that assess the functional genomics (the activity of genes) such as gene mapping, sequencing, statistical genetics, and expression analysis (detecting levels of protein synthesis), are applied in an attempt to isolate a gene locus that is associated with a differential physiological response in the presence of the drug.[8] Alternatively, these studies can be done in the absence of a drug to compare a sample from a diseased individual with a sample from a healthy person to hopefully uncover markers of disease, which may then be useful as drug targets. Databases with information ranging from DNA sequences to gene expression patterns are combined with powerful bioinformatic tools to identify the specific genes involved with disease and which variations interact with the drugs in question.

Using genetic information to predict drug response is not as simple as the explanation above might suggest. If medicine becomes personalized, then, by logical extension, it may seem that each patient's diagnosis will be unique. Because individuals have unique genomes, it may seem that to personalize medicine, the necessary strategy would be to test every therapy

---

[8]Regalado A. Inventing the Pharmacogenomics Business. *Am. J. Health. Syst. Pharm.*, *56*, 40–50, 1999.

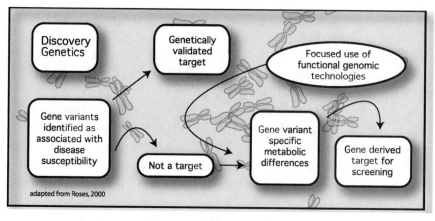

(a)

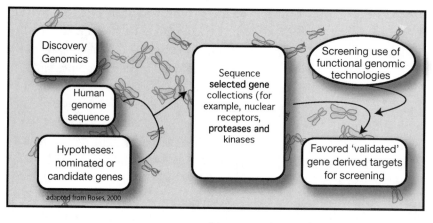

(b)

Figure 4-1 Using genetics (a) and genomics (b) to identify drug targets. Reprinted with permission from *Nature, 405,* 857–865, 2000.

against every genome. However, two factors obviate this experimentally impossible prospect. First, although it is true that everyone has a unique genetic code, most people share the majority (above 99.9%) of their DNA sequences. The remaining variations are substantial (considering that 0.1% variation is equivalent to roughly 3,000,000 bases), but the number of genes in question is limited to those involved in disease or metabolizing drugs.

Second, there are only a limited number of genetic variations that have survived evolution, and they are being mapped by a number of organizations to ease this interpretive hurdle. For instance, the SNP Consortium has more than 1,500,000 of these variations (SNPs) in validation at the time of this writing.[9] Aware of the importance of correlating this information to everything from specific disease states to drug response, the group has formed alliances with companies like Orchid Biocomputer, which has the experience and interest to both confirm the accuracy of SNP data and classify it according to clinically significant variations.

The significance of SNPs cannot be overstated; however, they require a great deal of analysis, categorization, and interpretation to aid clinical diagnosis. By definition, SNPs are "single-base differences in the DNA sequence that can be observed between individuals in the population."[10] The important clause in this description is that they can be observed between individuals in the population, meaning that many share these variations. More importantly, these genetic similarities imply phenotypic similarities; however, one SNP does not always correspond to one phenotype, making it difficult to derive data from a single variation. One SNP in a disease susceptibility gene may occur in many people, but only a few may have the disease or drug response profile thought to be associated with the SNP. This suggests that other factors may be involved that interact with the gene containing the SNP, factors that are important for the full phenotype to be expressed. Interestingly, those few that share phenotypes may do so with their relatives, implying a pattern of genetic inheritance that involves multiple SNPs acting in concert. Scientists have confronted this phenomenon by looking at multiple SNPs that are close together on a single chromosome. For instance, an individual may inherit two SNPs that influence a disease or response to a specific drug, but one originates from the chromosome donated by the father and the other from the one donated by the mother. They are not linked on a single chromosome in a single gene, so there is no clinical significance. That is, the gene must vary from the normal version in at least two regions per gene copy to have a clinically significant impact. It is important that these varia-

---

[9] http://snp.cshl.org.
[10] Roses, Allen D. Nature, 405, 861, 2000.

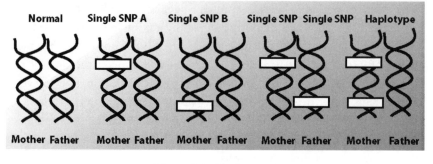

Figure 4-2 Haplotypes.

tions be on a single chromosome in such cases because the molecular machinery that translates DNA into proteins will read one stretch of DNA, incorporating the two differences into a single gene product. Known as linkage disequillibrium (a statistical term indicating that the association of these variations occurs considerably more frequently than random probability would suggest), frequently occurring associations of SNPs on a single chromosome (haplotypes) may offer a more accurate mode of analysis when compared to singular SNP analysis. (See Figure 4-2.)

In a well-known example of this phenomenon, the beta-adrenergic receptor (β_2-AR) gene consists of different forms, and one haplotype in its sequence correlates to poor responsiveness to the asthma drug albuterol. Although there are a number of SNPs that work alone to create different gene products, historically, many of them have only been linked probabilistically; however, when two specific SNPs were analyzed together, they were found to confer clinically significant alterations in the gene product. Working in tandem, these SNPs cause the gene's transcribed protein to change its form substantially enough to prohibit albuterol from binding to this protein. Albuterol, technically known as a beta-adrenergic agonist, has no effect on individuals who carry two copies of this haplotype version of β_2-AR;[11] that is,

[11]Lima, J. J. et al. Impact of Genetic Polymorphisms of the β_2-Adrenergic Receptor on Albuterol Bronchodilator Pharmacodynamics. *Clin. Pharmacol. Ther.*, 65, 519–525, 1999.

individuals whose mother and father each passed on this specific version of the gene are homozygous for these two SNPs in linkage disequilibrium.

To state that functional changes in genetics only occur in haplotype form would be inaccurate. In many cases, SNPs are in linkage disequilibrium but only one of these SNPs is functionally significant. However, analyzing patterns of inheritance has revealed that many functional SNPs are carried over in concert with a nonfunctional variant. One might wonder why the nonfunctional variant (e.g., it may occur in a noncoding region of the gene) would be of interest. There are at least two very important reasons why. The first is as an aid for analysis. Personalized medicine relies on a number of tools, ranging from functional genetics to pharmacology. One critical tool, although sometimes undervalued, is statistical genetics, which brings together the fields of statistics, population analysis, epidemiology, evolutionary biology, and genetics. Bioinformatics, the application of analytical software tools to biological analysis, has subsumed this discipline, which may explain why its contributions are less apparent than those of other disciplines. However, without statistical genetics, no one would know how to use this information on a large scale. SNPs are so variable across the population that personalizing medicine according to an individual's SNP profile would result in so much statistical noise that the information would be almost useless. Combining SNPs into haplotypes significantly reduces this noise by classifying individuals according to their commonalities. With haplotypes, taxonomical trees can be structured in which genetic branches correspond to particular drugs or doses.

The second reason for grouping SNPs according to haplotypes centers on our expanding knowledge base. Because we are still learning about the genome, there is a chance that intron regions (the noncoding regions of a gene that separate those areas that are translated into a protein) and other, presumably unimportant, DNA sequences might have greater functional significance than previously thought. Ignoring associations that have survived thousands of years of evolution runs the risk of neglecting a significant phenomenon for which the Rosetta stone has not yet been discovered. An example might be the *ApoE* gene, which is linked to Alzheimer's disease. In this case, a haplotype version of the gene, known as the *ApoE ε* 4 variant, can help predict susceptibility to Alzheimer's, as well as response to Tacrine, a promising treatment for the disease.[12] Interestingly, this gene codes for apolipoprotein, which facilitates the binding of lipoprotein (cholesterol) to receptors on cells. Although it may seem that this link strongly implicates the *ApoE* gene in Alzheimer's pathology, compelling biological evidence for this association is only now being discovered, a good six years after the association between the gene and drug response was reported.

Personalized medicine may be the most significant contribution derived from knowledge of the human genome. Although it has always been medicine's goal to tailor treatments to individual needs, technological and economic restraints have traditionally facilitated the development of therapies according to a "one size fits all" strategy. Drugs were targeted at the highest percentage of the population that would respond, but current technologies can change this approach. Soon, the melding of genetics, pharmacology, and bioinformatics will yield a new era in patient care, in which medicine will finally be delivered according to personalized parameters, as it should be.

III. The Benefits

From the name itself, the benefits of personalized medicine may seem fairly obvious; however, they are farther reaching than a first glance might suggest. Ultimately, all progress in this field will filter down to the patient, but it offers more advantages than better healthcare. Pharmacogenetic and pharmacogenomic drug development strategies will also yield significant advantages to the pharmaceutical industry. Essentially, there are three levels at which this medical revolution will improve care giving. The life science industry will develop treatments more efficiently, while the end consumer, that is anyone who needs pharmaceuticals, will have dramatically improved healthcare. Finally, treatments can be targeted more effectively, decreasing adverse events while improving the overall efficacy of therapy, thus decreasing the costs to society for delivering healthcare.

As mentioned in the previous chapter, developing a drug is a very expensive, time-consuming proposition. Worse still, each attempt is similar to

---

[12]Poirier J. et al. Apolipoprotein E4 Allele as a Predictor of Cholinergic Deficits and Treatment Outcome in Alzheimer's disease. *PNAS USA, 92,* 12260–12264, 1995.

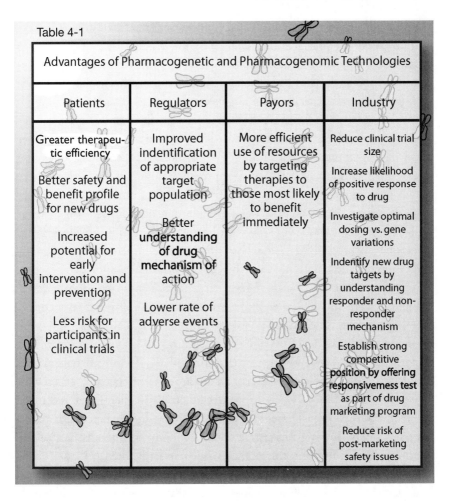

Table 4-1

Advantages of Pharmacogenetic and Pharmacogenomic Technologies			
Patients	Regulators	Payors	Industry
Greater therapeutic efficiency Better safety and benefit profile for new drugs Increased potential for early intervention and prevention Less risk for participants in clinical trials	Improved indentification of appropriate target population Better **understanding of drug mechanism of** action Lower rate of adverse events	More efficient use of resources by targeting therapies to those most likely to benefit immediately	Reduce clinical trial size Increase likelihood of positive response to drug Investigate optimal dosing vs. gene variations Indentify new drug targets by understanding responder and non-responder mechanism Establish strong competitive **position by offering responsivemess test** as part of drug marketing program Reduce risk of post-marketing safety issues

making a bet with odds that would cause most gamblers to cringe. Taking between 12–15 years on average, the one of hundreds of drugs that makes it to market may cost up to $500 million to get there. To recoup fully its expense, and those of the failed attempts, it *has* to be a "blockbuster" drug; that is, it should approach the success of a rare drug like Prozac, demand for which has lead to revenues in excess of billions of dollars a year.

Pharmacogenetics/pharmacogenomics are discovery methodologies that can dramatically decrease the time and costs currently involved with obtaining drug approval from the FDA. Facilitating each aspect from target selection to safety testing, melding genetics with pharmacology promises to streamline the usually cumbersome process. This technology platform would aid the analysis of the genetic basis of disease. After isolating disease susceptibility genes, specific drug targets can be chosen based on this knowledge. Currently, most drugs are developed irrationally; that is, large numbers of molecules are created blindly, based on fragmented hypotheses. After hundreds or thousands of different molecules are produced, they are screened by traditional pharmacological methodologies in hopes that one of these agents might meet the basic criteria necessary to move on to the next stage of testing. With pharmacogenetics/pharmacogenomics, drugs are rationally designed to target the specific disease pathway implicated by genetic variations in susceptibility genes. Essentially, novel drug targets are identified faster and more logically than the current brute force approach.

The elegance of the pharmacogenetics/genomics approach to drug development extends much further into the process. When drugs are developed, they must pass through three phases according to regulations set down by the Food and Drug Administration.[13] According to Allen Roses, "medicine response profiles could be identified during phase II clinical trials. These could be used in the selection of patient groups enriched for efficacy in phase III studies. This is likely to make these trials smaller, faster and more efficient."[14] In the first phase, following discovery, the therapeutic is tested for basic safety. During phase II a small sampling of humans receive the treatment, and if it passes this phase's strict standards, then it may proceed to phase III testing. In the third phase, the number of people included in the trial increases dramatically, and relevant information discovered in the previous phase is included in this succeeding protocol. As Roses claims, some of the phase II information can account for variable response if genetics are analyzed alongside drug response. With this information, the population for testing in the subsequent phase can be chosen based on genetically predicted drug response. Including this information in the protocol would not only reduce the size of the testing population compared to current trials, but sub-

---

[13] *http://www.fda.gov.*
[14] Roses, Allen D. *Nature, 405,* 862, 2000.

stantially improve the response rate over a randomly chosen population. A smaller, rationally chosen population means a cheaper and swifter clinical trials process.

These experiments will do more than characterize the genetic basis of drug response; they will also help explain genetically mediated adverse drug reactions (ADR) and side effects. Although these may not seem profound to most, anyone who has suffered a severe reaction to a pharmaceutical knows how painful and dangerous such an event can be. In fact, in a recent study published in the *Journal of the American Medical Association* (*JAMA*), researchers concluded that in one year, over 2 million people were hospitalized with serious ADRs and over 100,000 of these cases were fatal, making ADRs between the fourth and sixth leading cause of death in the United States.[15]

Currently, individuals who suffer from acute lymphoblastic leukemia (ALL), a severe cancer affecting children in their teens and younger, are prescribed the chemotherapy drug 6-mercaptopurine. For those who can adequately metabolize this drug, there is greater hope for survival and remission; however, some people have a genetically determined impairment in the molecular machinery that metabolizes this drug. Until recently, poor responders suffered from more than the symptoms of ALL; they also felt the toxic effects of building up the drug in their bodies. Fortunately, this outcome can now be avoided since it was discovered that a gene known as *TMPT* exists as a polymorphic variant that creates the "wrong" version of the metabolizing protein, thiopurine methyltransferase.[16] This alternate version of the protein cannot break down all of the molecules of the drug into their functional form, which not only defeats treatment, but also endangers the patient. A new test for the *TPMT* variant enables doctors to prescribe clinically effective doses, which helps the patients achieve health benefits without the dangerous side effects that might occur due to a *TPMT* mutation.

Of the benefits listed above, none compare to the ultimate goal of this technology: personalzing medicine. To make medicine personal requires a

---

[15]Lazarou, J. et al. Incidence of Adverse Drug Reactions in Hospitalized Patients: A Meta-analysis of Prospective Studies. *JAMA, 279,* 1200–1205, 1998.
[16]Lennard L. et al. Genetic Variation in Response to 6-mercaptopurine for Childhood Acute Lymphoblastic Leukaemia. *Lancet, 28,* 336 (8709), 225–229, 1990.

series of experiments during the drug development process. Ideally, information will be accumulated during clinical trials. While observing how patients respond to drugs, genes associated with the drug target can be examined for SNPs and haplotypes. After categorizing patients according to drug response, the genotypes associated with each level of response can be used in the future to determine whether a patient should receive a particular treatment, as well as which dosage. Information collected and disseminated by computers will match each drug to a response profile based on genetics. Complex genetic interactions will become an aid to medicine, rather than the detriment that had previously challenged patient care.

Personalized medicine may seem an unoriginal goal of medicine, but realizing it will be a revolutionary shift. The idea of individualized care has always been the basis of doctor/patient relationships, but when it comes to pharmaceutical therapy, there has always been quite a bit of guesswork and

Figure 4-3.

experimentation involved with each patient. Going back a few steps, to the pharmaceutical company developing a drug, it has always been the goal to create a remedy that targeted the majority of a diseased population, rather than developing multiple drugs that would address them all. The associated costs and resources demanded such a strategy. If the marketed product exhibited efficacy in 80% of the trial population, then it would be considered an excellent drug. After that drug made it to market, a doctor might prescribe it to a patient who could take it for months with no effect. Following the failure of one drug, another would take its place, and the cycle would be repeated until the best match between patient and drug was finally discovered.

With pharmacogenetics, a drug's efficacy of response for an individual will be known prior to the patient taking it. One might wonder why this incarnation of personalized medicine is revolutionary, and an unreflective patient may even think this is the current state of medicine. Yet this "one size fits all" strategy is poised to be overturned. In an almost Jeffersonian fashion, the greatest good for the greatest number has prevailed, but now the greatest good is for all, not just the majority. As Roses explains, "the application of pharmacogenetics will not diminish the population in whom [sic] a drug is effective, but simply allow prediction of patient response rather than prolonged and expensive prescribing by trial and error."[17] When pharmacogenetics reaches its ultimate promise, every patient in a doctor's office will be prescribed remedies that are unique to his or her genetic structure. It will no longer be a game of odds, hoping that the patient falls into that percentage of people who will respond to a particular therapy. A quick genotype derived from a drop of blood or saliva will be scanned into a computer that will diagnose and recommend a therapy that fits the genetic profile of the patient, optimized by drug and dosage.

While it may seem that only the patient and the pharmaceutical company stand to gain from this revolution in medicine, healthcare systems will also benefit. With such a precise mode of delivering care, the cost benefit of each therapy will be maximized. Insurance companies and HMOs will not spend money on therapies whose ineffectiveness leads to further complications of

---

[17]Roses, Allen. *Nature, 405,* 862, 2000.

disease; furthermore, avoiding adverse events will also decrease costs. All of these elements working in concert will aid the individuals in a healthcare system because the efficacy of precise treatment will obviate the need for rationing. Improved patient health leads to fewer hospital stays, emergency room visits, and other high-cost actions that draw on the resources of payers. There are very few who stand to lose in this equation; however, research efforts must proceed far and freely enough for the necessary tools to make it into the clinic. At present, the shift may be ten years away, and the owners of the information are split between industry and the public sector. For this revolution to be fully realized, the information has to make it into the public domain, and any group that restricts it will cause a bottleneck that prevents a remarkable transformation in healthcare.

IV. The Issues

Personalized medicine, despite its unique nature, does not have entirely unique ethical concerns. In fact, amongst the most important factors in this arena are the exact issues explored in the previous chapter on DNA data banking: privacy, confidentiality, and, by extension, informed consent. Unique genetic information will be collected from each patient, and this information can be misused by employers and insurance companies. Furthermore, the implications for ethnic groups resurface. For this reason, to thoroughly understand the bioethical concerns surrounding this field, revisiting the discussion on data banking in Chapter 3 is critical.

One implication that was touched on briefly in the previous chapter and deserves attention both there and here is the psychological impact of genetic information on a patient. Speaking to a geneticist or another specialist close to this field, it may be difficult to convince him or her of the difference between genetic information and other biomarkers like cholesterol level or blood pressure. Both of these examples are risk factors for cardiovascular disease, and if one considers how they are used to assess patient health, a very precise equation for risk stratification emerges. Certain risk factors are weighted more heavily than others, and at the end of the assessment, a patient is told his or her likelihood of getting the disease. Afterwards, a particular course of action, perhaps a combination of drugs, diet, and exercise, is prescribed.

Differing from biomarkers, a person's genetic code does not change. While these markers are dynamic—cholesterol can go up and down—a person's genes are static. Because they do not change, genes and gene variants dramatically increase diagnostic precision. Biomarkers may indicate a predisposition or a likelihood of disease, but a better understanding of genetics may shift diagnoses from approximations to certainties. Although a tremendous boon to medicine and patient care, genetic information may also inflict psychological and other injuries on the patient. In his analysis of the issues surrounding pharmacogenetics, bioethicist John A. Robertson explains these potential problems:

> A person's knowledge of variations in his genotype may radically change his self-image and his life and his reproductive plans. If preventive or ameliorative action is possible at all, it may be highly intrusive and not covered by health insurance. If other persons become aware of the information, they might stigmatize the person or use it adversely in insurance or employment decisions.[18]

Although the issues surrounding employability and insurability were duly addressed in Chapter 3, the psychological concern and the alteration of life choice that may follow genetic testing require attention in this part of the discussion. Because of the novelty surrounding genetic information, individuals may have difficulty coping with immutable conceptions about their health. When a seemingly healthy patient visits her doctor, she does so under the impression that medical problems can be overcome with some type of intervention. Learning that her genes prevents her from using a drug that others can normally metabolize can cause anxiety. DNA has taken such a prominent symbolic place in our society that one cannot help but feel that it plays an integral role in our individuality; thus, variations in one's genetic code may not only cause the incomplete metabolization of a drug, it may

---

[18]Robertson, John A. Consent and Privacy in Pharmacogenetics Testing. *Nature Genetics, 28,* 207, 2001.

leave a patient feeling incomplete altogether. In cases where a symbolic conception of DNA threatens its medical utility, clinicians and genetic counselors need to convey the important balance between genetics and the environment. Treatments are not natural; indeed, they are human artifacts, and when placed in the body, they are environmental influences. Everyone, both physiologically and cognitively, reacts to the environment differently. Indeed, we are all individuals who experience the entire world differently; what could be more natural than responding poorly to one environmental factor in favor of another, even in the case of therapeutics?

Related to this issue lies the unlikely, but important, scenario in which pharmacogenetics may remove singular treatments from a disease population. Consider the following hypothetical scenario. A drug that has been on the market for a disease that is difficult to cure or even treat—imagine that a drug was developed for the dementia-causing Creutzfeldt–Jakob disease for instance—has been on the market for some time without pharmacogenetics/genomic analysis. Most patients who take the drug experience some degree of improvement in cognitive functioning, and the drug developer decides that it is time to improve efficacy via pharmacogenetic analysis. The firm hopes that the analysis will yield better dosing data, but instead, it is left with a more profound discovery. After performing such experimentation, an investigator discovers that a series of genes are linked to an all-or-nothing response profile; either a patient will respond to this new drug, or she will not. Through this discovery, scientists may have removed one population from the only treatment option that gives hope to the victim and his or her family. Although the placebo effect of this hypothetical drug may be powerful, its efficacy is questionable. Although an unlikely scenario, the implications of developing pharmacogenetic tests around singular therapies ought to be considered before proceeding. It may be better to proceed with such an analysis only if alternative treatment options of comparable efficacy are available.

Although critiques that center on the "symbolic value" of DNA and other more humanistic strategies may seem too esoteric to be practical, they are nonetheless indicative of the social resistance that companies will face upon adopting pharmacogenetic/genomic technologies. In a similar vein, the concept of linguistic obstacles is also pressing. Symbolic value is not only accorded to DNA; the term "genetic," particularly when followed by the word "testing," also carries substantial social stigmas. Perhaps a more relevant way of putting it is that the phrase "genetic testing" has many preconceived notions linked to it. Usually associated with revealing disease genes and susceptibility to inherited disorders, many are reluctant to be genetically tested for fear of discovering a "preprogrammed" disease. Connotations mirroring the loss of employment and insurability discussed in Chapter 3 understandably discourage people from undergoing these analyses. However, pharmacogenetic/genomic testing does not reveal disease information, although it someday may. The difference lies in the state of the patient upon such analysis, and, more importantly, the type of information that it may reveal. In most cases, patients will already have a disease, and the examination will only be used to choose a course of therapy. The goal is to refine treatment, not discover disease, a distinction aptly noted by Orchid Bioscience: "Pharmacogenetics is not about disease diagnosis, but rather drug efficacy and drug safety. . . ."[19] Although the ultimate aim of pharmacogenetics is to work in tandem with genetic testing to reveal disease and sponsor treatments prior to the manifestation of any symptoms, this will not become a reality until treatments for genetically mediated diseases are developed. Thus, the standard concern of diagnosing untreatable diseases will be obviated. However, should the current pattern continue, where diagnostics emerge that focus on diseases as of yet untreatable, a company would have to carefully consider the social value of releasing such a tool into a clinic. Regardless of either scenario, any group pursuing such research must make the distinction between genetic testing and pharmacogenetic testing.

Less esoteric than arguments over symbolism and linguistics are the economic impediments to this research. In the other chapters of this text, all of the technologies are framed in terms of bioethics, but within pharmacogenetics lay substantial questions of business ethics. To some companies, hearing that personalized medicine is in the *future* of healthcare suits them just fine, so long as it is not the immediate future. Big Pharma recognizes pharmacogenetics as a threat to many of their blockbuster drugs; that is, they fear

[19]Barash, Carol I. and Grant, Denis. *Potential Ethical Issues in Phramacogenetics.* Princeton, NJ: Orchid Biosciences.

that their pharmaceuticals that sell in the billions of dollars a year may lose market share if personalized medicine proves that some people currently on their drug should be excluded from treatment. Until pharmacogenetics alters it, the prevailing paradigm for the pharmaceutical industry is to develop the one drug that will penetrate an entire disease demographic, regardless of efficacy. Not a single drug on the market can treat every patient suffering from its targeted disease, due in no small part to genetic variation; thus, marketing efforts replace efficacy challenges. Profit replaces public health motives, since completely satisfying the latter is impossible. In these cases, when a drug has successfully captured the market, no drug company wants to be the first to limit its market share by restricting drug access to just a subpopulation of its current consumer base. For instance, a drug may be reaching 80% of the asthmatic population but is only truly effective in 50% of the population; pharmacogenetics might provide a simple test to reveal that only 50% of a company's consumers respond to it, and the loss of 50% of current customers could be crippling enough to disregard any options that contribute to a decreased market share.

Pharmaceutical companies greet pharmacogenetics with both trepidation and enthusiasm. This paradoxical approach offers insight into the pharmaceutical industry's strategy; Big Pharma recognizes that an inevitable shift will occur in the direction of personalized medicine, but they will not be a party to it until they have extracted as much profit from current methodologies as possible. Quoted frequently in this chapter because he is considered by many experts to be the dean of pharmacogenetics, Allen Roses explains that pharmacogenetics "is a disruptive technology, not a technology that sustains what organizations are used to doing. It's going to be a part of everybody's business, and that's what most people don't seem to understand."[20] Ironically, in a recent *Wall Street Journal* article, Roses' allegiance to the pharmaceutical industry trumps his enthusiasm for the social benefits of the technology he helped pioneer; in the article, when asked about other companies developing drug response tests for pharmaceuticals produced by his own firm, GlaxoSmithKline, (even though Glaxo has no intention of developing a response test on its own) he answers, "They can go screw with someone else's drugs."[21] Taken out

---

[20]Wortman, Marc. Medicine Gets Personal. *Technology Review,* January/February, 2001.

of context, this answer may seem unnecessarily harsh; however, it is intended to show that part of the disruption associated with pharmacogenetics stems from the potential challenges that drug makers face from genetic analysis companies. Biotech companies can leverage genetic information to develop analytic tools that challenge the market share of blockbuster drugs by correlating drug response to SNP variations. Such a development severely threatens the expected return on investment that motivates many pharmaceutical companies to develop therapeutics. However, if this tension remains unresolved, the greatest loser in this struggle will be neither the pharmaceutical industry nor the upstart biotech companies; rather, it will be the patient. Unfortunately, we are all, to some extent, the patient.

Another ethical concern facing personalized medicine reflects the growing divide of healthcare resource allocation between developed and developing countries. Unfortunately, the relationship between race and socioeconomic status in the United States may well contribute to this growing chasm if pharmacogenetics researchers proceed with experimentation that does not account for racial difference. Because ethnic variations correspond to important genetic variations for drug response profiles, racially homogenous clinical trials may limit more than scientific knowledge; it may damage individuals in developing nations whose need for novel therapies are as great, if not greater, than those who will readily access these treatments. As pharmacogenomic experts Howard L. McLeod and William E. Evans explain,

> Although the promise of pharmacogenomics is enormous, it is likely to have the greatest initial benefit for patients in developed countries, owing to expense, availability of technology and the focus of initial research. However, pharmacogenomics should ultimately be useful to world populations. There is clear evidence of ethnic variation in disease risk, disease incidence, and response to therapy. In addition, many polymorphic drug metabolizing enzymes have qualitative and quantitative differences among racial groups.[22]

---

[21]Anand, Geeta. Big Drug Companies Try to Delay "Personalized" Medicine Regimens. *The Wall Street Journal,* June 18, 2001.

[22]McLeod, Howard L. and Evans, William E. Pharmacogenomics: Unlocking the Human Genome for Better Drug Therapy. *Ann. Rev. Pharmacol. Toxicol., 41,* 114, 2001.

Excluding certain racial groups from pharmacogenetic trials runs the very real risk of creating a technical barrier to delivering therapies to the Third World, which may be added to the current political, social and economic barriers.

Using the United States as an example, a scenario may play out in which drugs are developed almost exclusively for limited, traditionally wealthy populations. The benefits of performing genetic analysis alongside drug testing may very well reinforce this problem. For instance, in the future the FDA may adopt the recommendations of pharmacogenetic experts and require that all drug development be accompanied by genetic analysis to predict drug response. If a predominantly Caucasian population dominates a clinical trial, a drug response profile may emerge that fits only this demographic. The drug can proceed through development, and its safety and efficacy may be validated based on this cohort. When it is finally released to the public, it will be done with a required pharmacogenetic analysis, perhaps a DNA microarray, which contains a panel of SNPs identified during the trial as predictors of differential drug response. Because both the SNPs and the drug itself were developed and approved based on a specialized population, it would be as if the "personalized" treatment were instead "ethnicized." Side effect and efficacy data could be so specialized that it would be unsafe to release it to a population for whom a pharmacogenetic analysis was never performed. If the cohort did not contain enough Africans, Chinese, Indians, and other ethnic groups that both represent a significant proportion of the world population and inhabit the majority of underprivileged countries, then the current debate surrounding delivering care to these countries may become scientifically moot. Politicians may no longer claim that economic infeasibility and cultural incompatibility preclude the distribution of therapies to the Third World; instead, they may claim scientific impossibility. After all, opponents might ask, how ethical would it be to send medicines to a developing country when the so-called cures have never been approved for use on that populace?

The severity of this approach can be crippling to developing countries. Traditionally, these nations have been below most life science companies' radar screens, and when they are in view, it is because human rights groups have forced the issue. Speaking candidly on the topic, Dr. Bernard Pecoul of Doctors Without Borders explains that economic pressures influence too much of the drug development process, limiting both the types of drugs developed and their geographic distribution: "Pharmaceutical companies will always aim for maximum profits by marketing a new obesity drug rather than pioneering a novel malaria treatment. . . . When new vaccines or medicines are developed, most of the world's population is left out of the picture."[23] If such a trend continues with pharmacogenetics, the barriers to global health will reach entirely new levels of complexity.

Considering the current state of world health, complicating matters may seem difficult, but pharmacogenetics offers tremendous hope if it instead is used to simplify them. Pharmaceutical companies can make a substantial difference today if they rally behind the cause of world health. Nothing exemplifies this more than the current status of AZT, an AIDS treatment, in Africa. Recently, this drug has been released to Africa at a substantially discounted rate. However, the battle to get it there waged on for years. Drug companies were reluctant to release generic versions or name brand versions of these drugs, claiming their unregulated distribution would only result in increased black market trading. Critics countered by condemning big business' audacity at asserting cultural expertise over a foreign populace while actively impeding public health. Very active in these debates, the World Health Organization (WHO) has endeavored to mediate the shaky ground that both factions tread to find a solution to alleviate the suffering of underprivileged populations that cannot access conventional therapies. As evidenced by the WHO's focus on problems like the AIDS epidemic in Africa, drugs that can help a population do not always make it to the areas where the need is greatest because political, economic, and industrial motives interfere with public health. However, when pharmaceutical companies do make the effort to aid populations by decreasing prices or other initiatives, dramatic changes ensue. In a recent statement, Dr. Gro Harlem Brundtland, Director-General of the World Health Organization, extolled the virtues of such efforts:

We health professionals must now do even more to confront the global HIV/AIDS epidemic and respond to its consequences. . . . Pharma-

---

[23]McNeil, Donald G., Jr. Drug Companies and the Third World: A Case Study in Neglect. *The New York Times,* May 21, 2000.

ceutical companies have made dramatic reductions in the prices of medicines necessary for treating people affected by HIV, thus enabling better access to means for preventing mother-to-child transmission of the virus in the least developed nations. This shows how better access to effective AIDS care and successful prevention of HIV infection are inextricably linked.[24]

This is an important victory on the road to better health; however, how might this statement have changed if the previously resistant pharmaceutical companies could claim that there was just not enough data to support the efficacy of treatment in these populations?

There are some barriers to overcoming the socioeconomic–racial divide that may grow due to pharmacogenetics. The threat of biopiracy, the concept of "stealing" DNA from cultural groups, has made many groups in less-developed countries wary of contributing their DNA to companies and foreign researchers. With potential violations of privacy, confidentiality, and respect fueling this fire, a delicate balance must be struck to protect these interests. Convincing resistant groups and activists that genetic material from all ethnic groups is as critical to the donors as it is to the drug companies is a difficult task, but it is vital that these experiments be as inclusive as possible. Another problem lies with defining racial groups. Although we recognize skin color as a parameter, genetics do not correspond to these cultural precepts. Certain genetic markers will do better than the social distinctions currently used; however, these markers have yet to be identified, despite the great promise such a method offers.

A final note on the ethics of pharmacogenetics touches upon one of the older concerns facing genetic technologies, which is the patenting of genes. Controversy over this topic places businesses in a very difficult position, especially because businesses *have* to patent genes to survive. More precisely, and a critical distinction in this argument, companies patent the use of genes, not the genes themselves; that is, firms determine how they may use the genes for applications such as diagnostics or drug discovery. Industrial competition demands that companies developing products based on novel

---

[24]*http://www.who.int/inf-pr-2001/en/state2001-13.html.*

genes must patent them. Although patenting a gene may intuitively seem to violate social interests, the process is meant to strike a balance between social and industrial concerns, as bioethicists Timothy Caulfield, E. Richard Gold, and Mildred K. Cho explain,

> Patents are designed to provide incentives for innovation and for the development of products from which the public can benefit. The logic behind patents is straightforward. If we pay people to invent, produce and distribute products (by granting them limited monopolies), they will do so. Unfortunately, this logic does not always work.[25]

The spiral began when groups spent a vast amount of time and money on discovering genes. To recoup their costs and appease both investors and Wall Street, firms had to patent their discoveries. Intellectual property adds to a company's assets, something that was difficult to increase in the early years of biotechnology. After this trend took hold, it could not easily be over-turned. This explanation does not constitute a justification, but it does out-line the necessity of patents. Without protection over investments of time, money or manpower, none of the cures that the world depends on could or would be developed. In terms of capital, a firm can spend millions of dollars, indeed its entire operating budget, in search of a gene. If it is not patented soon after discovery, then another company with more capital could develop and market products based on the discovery. The group that both found the gene and established its functionality (genes can only be patented when their function is known, thus no one can patent sequences of DNA anticipating that they will some day be important) might very well go bankrupt while another comes along and exploits its hard work.

Those opposed to patenting genes care little for the practical concerns of industry. Their arguments center on both the commodification of nature and the limitations facing research when one person or group "owns" a gene. With regard to the first point, most people feel uneasy when something that exists within them, like a gene, becomes the property of another party. It was

---

[25]Caulfield, T. Gold, E. R., and Cho, M. K. Patenting Human Genetic Material: Refocusing the Debate. *Nature Reviews Genetics, 1,* 229, December, 2000.

not, after all, "invented." Nature or God (or both, depending upon one's religious views) did all the work, and technology merely revealed a pattern that has existed for millennia. Furthermore, to reduce something so grand as humanity's biology to a legal and economic issue seems to cheapen the concept of our biology. In many people's eyes, our DNA should transcend both legal jargon and dollar signs.

Whether patents limit research or stimulate it can be debated over and over, but patents do not restrict not-for-profit research. The concern that intellectual property might destroy the research perogative mistakes the function of patents. Patents only limit making profit off of the novelty of someone else's work. To complain that patents stop academic research is a hidden way of masking animosity toward the private sector. Anyone can use any patented invention freely, so long as no one profits by it. University labs can pursue research related to patented information, but if coming up with their own intellectual property and royalties motivates them, then they are obviously not pursuing research for the sake of altruism. (It is interesting to note that the University of California has one of the largest intellectual property portfolios related to biotechnology in the world.) Similarly, companies can use patented information and techniques in-house for their own investigations; they simply cannot develop products around another group's intellectual property (IP, another phrase describing a group's patent position) without working out a licensing deal. Patents do not impede research, but profits may. Patents do not inherently restrict research; they only define the distribution of capital derived from an invention. One company may have all of the patents on a gene, and therefore earn fees from anyone who uses the IP for revenue. However, if four companies own an equal number of patents on a gene (one may own testing rights, another transcription rights, etc.), then each has to share the revenues on the gene according to their patents.

V. Industry

Because the necessary technologies to introduce pharmacogenetics into public health have only emerged over the past few years, personalized medicine has witnessed very few applications. As such, the above discussion of the technology's benefits and issues are speculative, but the unique convergence of these factors have inspired many companies to pursue this research. Almost all of the major players in the life science industry recognize that personalized medicine is as much an inevitability as it is an imperative. As Gilbert Ruaño, CEO of pharmacogenetic company Genaissance Pharmaceuticals, states, "This is not a fad. . . . It's a major tidal wave changing the entire pattern of health care."[26]

Genaissance represents just one of many companies that plans to herald personalized medicine. To reach its technical milestones, it has completed very important studies that will frame its future scientific efforts. In a recent article in *Science,* Genaissance scientists published the results of a study analyzing 313 genes. According to their findings, these genes exist, on average, in fourteen different variants, with actual numbers ranging from two to fifty-three different versions of any given gene.[27] Variagenics is pursuing similar research, and both companies are betting that they will discover genes and gene variants that will lead to faster drug development and response profiles for new and available drugs.

Pharmaceutical companies are in a similar position, but they already have drugs in development, as well as those on the market, which will receive the same treatment. However, it remains to be seen how they will apply their discoveries. While Big Pharma could develop testing panels to indicate drug response for their marketed drugs, they may limit their research to identifying side effects rather than differential response. Drugs that already have a market may be economically compromised if research reveals that consumers ought to be specifically rather than generally targeted. Losing nonresponders is not a welcome option; however, side effects ought to be avoided at all costs, and drug companies do not want any patients reacting negatively to a therapy. Therefore, many pharmacogenetics analyses will focus on averting adverse events according to genetic information. Newer drugs will most likely receive total pharmacogenetic analysis, accounting for response dose and side effects, because they will be developed in tandem with other

---

[26]Wortman, Marc. Medicine Gets Personal. *Technology Review.* January/February, 2001.
[27]Stephens, J. C. et al. Haplotype Variation and Linkage Disequilibrium in 313 Human Genes. *Science, 293,* 489–493, July 20, 2001.

treatments. For instance, rather than develop a drug for osteoarthritis, a pharmaceutical company may know that three different genes affect the pathology of the disease. Therefore, three different treatments may be developed to target everyone in this disease population. The cost and speed of development will improve because of genetically targeting the trial population, while the target market remains the same. All osteoarthritics may be customers, but they will be prescribed one of the firm's three treatments rather than a single one that may not work.

To fully investigate how a company contends with the ethical issues involved with pharmacogenetic research, Interleukin Genetics will be analyzed because it focuses on just a few genes that are implicated in a host of diseases. At first a testing company, Interleukin's move into pharmacogenetics reflects its belief that the future of healthcare will focus on personalized medicine. The company built its technology platform around the interleukin-1 (IL-1) gene cluster on human chromosome 2, which contains a number of genes that control the body's inflammatory response to a wide variety of stimuli (or challenges). Since most illnesses are characterized by the body's reaction to a stimulus, rather than the stimulus itself, these few genes are very important in understanding disease pathology. For instance, high cholesterol is dangerous, but only because it triggers a cascade of events, partially mediated by interleukin molecules (also known as cytokines), which destabilize plaques and constrict arteries, causing the actual physiological damage. Cytokines also mediate the "disease" response in illnesses like rheumatoid arthritis, osteoporosis, and many other diseases.

Because of its role in autoimmune disorders and other inflammatory-mediated diseases, many drug companies have spent time developing treatments based on the IL-1 gene cluster and similar genes encoding cytokines. However, because Interleukin Genetics began its research on these genes, the company's intellectual property position allows it to proceed with pharmacogenetic research that will be important to many treatments developed externally. Unlike Genaissance and Variagenics, Interleukin does not primarily discover and patent new genes and gene variants; rather, it looks at well-characterized diseases and correlates drug response to particular variations in the IL-1 gene cluster.

The corporate approach to bioethics is not an alien concept to Interleukin. In fact, the current CEO, Philip R. Reilly, a widely published bioethicist, joined the company in late 2000 after spending many years advising the company on the ethics of the firm's first product, a genetic test for susceptibility to periodontal disease. Aware of the concerns surrounding genetic testing, the company thought very carefully about how the information would be handled. Knowing that the IL-1 genes were involved in many more diseases than periodontal disease, cofounder Ken Kornman and others made sure that Interleukin Genetics would perform the testing and only release phenotypic data, rather than reveal genotypic data, which may have broader implications should future research correlate disease to particular gene variants. Clinical data regarding the test results, and nothing else, is accessible to both the patient and his or her periodontist.

This kind of concern and foresight is rooted in the experience of the officers of the company. No stranger to the bioethical terrain, Ken Kornman had seen first hand the problems associated with poor ethical decision making. To him, good conduct equals good business and good science. While this was always an intuitive notion, the practical implications became well articulated when he performed a study in Sri Lanka. Planning to assess the incidence of periodontal disease and its correlation to the interleukin-1 gene cluster, Dr. Kornman participated in an NIH-funded study to genotype a population of miners that had varying degrees of disease, but consistently poor dental care. Before going to Sri Lanka, Kornman spent a good deal of time with an Institutional Review Board developing as comprehensive a consent form as possible. He also made sure to find translators who would not only work with the investigators, but would also explain the consent form to illiterate individuals. With all of this taken into consideration, and all of the advanced planning on both the scientific and ethical fronts, the study merely ended up consuming time and money but resulting in nothing.

When the research team went to Sri Lanka, the goal was to find family members and genotype these individuals in hopes of correlating hereditary disease patterns to inherited gene variants. With all of the research team's planning, many of whom had never performed an international study, the consent process missed an important factor. Translators do not necessarily understand the culture, and when individuals were recruited for the study, they were asked to bring family members to participate as well. The investigators were not completely familiar with the cultural differences between the United States and Sri Lanka; the expansive definition of "family" in the latter included close friends, rather than just blood relatives. Thus, inherited patterns became difficult to recognize. This was discovered too far into the study to salvage anything, and the findings never amounted to anything, but Dr. Kornman's intuition that social components are critical to investigating scientific questions was confirmed. It was the type of mistake that he would not allow his future colleagues to make.

Philip R. Reilly recognized a similar imperative in biotechnology. Having spent the majority of his research career in the not-for-profit sector, Dr. Reilly's background as a lawyer and M.D. drew him to biotechnology as an ethical advisor. His past involvement with Interleukin Genetics revealed that pharmacogenetics represented an intersection of genetics, ethics, and industry in a unique manner, where his skills are ideally suited. "I am very fortunate to be at Interleukin. It isn't always clear how to leverage my professional background into a corporate setting. Having spent so much time focused on biotechnology as a clinical geneticist, a lawyer and an ethcicist, it seemed like too perfect an opportunity to pass up."

Interleukin Genetics' goal, as most life science companies claim, has always focused on improving healthcare. The company sees personalized medicine as a way to take a substantial leap toward better treatments and greater overall public health. With this goal in mind, it has set out to correlate drug response of available therapies to genetic variations for the many diseases affected by the IL-1 gene cluster. Although this may meet with opposition by the pharmaceutical companies that are developing these treatments, achieving the promise of improved healthcare becomes a stronger motivation than the oppositional interests of Big Pharma.

The choice of pharmacogenetics as the company's technology platform was informed as much by economic as it was by ethical concerns; after all, it is first and foremost a business. When the company refocused its efforts in mid 2000, it specifically chose to become a pharmacogenetic company because of management's belief that personalized medicine would deliver a substantial return on investment and benefit to patients. Because of this transition, the business model for the company is dependent on improving health in a much more profound way than the life science industry is accustomed to. No longer infected with the notion that one therapy or diagnostic is enough for health, Interleukin needed to develop assays that would facilitate treating everyone. Although most of these experiments are in their early stages, the linking of the company's success to its costumer's health is much stronger than in many other firms, and proportionately risky.

Because of this association with its prospective customers, Interleukin Genetics has almost assumed the role of a patient advocacy group. There are forces at play resisting the transition to personalized medicine,[28] but the company has placed itself in a position in which it will either deliver this technology to the public, or it will not succeed. The first step is to bring attention to the need for the technology and the immediate applications that would benefit from pharmacogenetics. As noted earlier, pharmacogenetics has met a great deal of resistance by drug companies because it might limit Big Pharma's immediate market share. Yet, if a means of improving healthcare is within Interleukin's power to deliver, then both its core values and business model demand that it do so. As such, the company believes that it should lead an educational initiative detailing pharmacogenetics' benefits.

Choosing to educate and choosing whom to educate are two different matters. As Chief Medical Officer (CMO) Kip Martha, M.D. explains:

> We need to get pharmacogenetics to the patient, but it will be difficult if not impossible for patients themselves to force this change in medicine. While there is a general consensus amongst experts that pharmacogenetics will eventually transform healthcare, most do not expect it to happen soon. However, we know that there are certain applications where the benefits can be delivered now. Specifically, these applications concern existing drugs that act through well-understood mechanisms but where widely variable responses to the drugs are known to exist. By focusing our attention now on understanding how common variations in the genes, which produce the targets for these drugs, affect a person's response, we can promptly take the first critical steps down the path towards bringing personalized medicine to patients. Therefore, the charge

---

[28]Annand, G. *Wall Street Journal,* June 18, 2001.

to pharmacogenetics will not begin with a comprehensive, sweeping change of the system, but rather it will start with a few specific test cases brought to the fore, which will force all the major stakeholders (regulatory authorities, payors, ethicists, lawyers, pharmaceutical companies, and patients) to grapple with these complex issues thereby setting precedents which pave the way for the widespread change to follow.

Whereas most people are saying that the science will not be applied for another ten years, Dr. Martha asks: Why not? "Clinical trials can start these analyses now, which would vastly reduce the amount of time and number of adverse events involved." Knowing this, the obvious choice for focusing educational efforts may seem to be on the pharmaceutical companies. However, these firms already realize the promise of personalized medicine, which is why they have in-house pharmacogenetics programs. Most likely, these programs are dedicated to testing for side effects to Big Pharma's current drugs, rather than response profiling. The former represents prescriptions that ought never be delivered if negative side effects are too great, whereas the latter could limit the marketability of a drug. In terms of response profiling, pharmaceutical companies will only release this type of information when they have a panel of drugs that treat a particular disease, rather than the single "blockbuster" strategy that they usually follow.

If the pharmaceutical companies already know the story, then another group has to be targeted. After weighing different options, Dr. Martha realized that the most concerned group officially representing the public interest in matters of drug development is the Food and Drug Administration (FDA). If any group needs to realize that pharmacogenetics can aid healthcare in the immediacy, it is the FDA. Granted, the agency is far from ignorant on the matter; however, the administration's responsibilities are often too vast to account for every medical contingency. Thus, when Interleukin Genetics' research program discovered a number of areas in which pharmacogenetics showed strong promise, ranging from drug development to its delivery, the company put together a presentation for the government agency.

In late July, 2001, Interleukin Genetics, represented by Dr. Reilly and Dr. Martha, visited Washington DC to promote the benefits of the technology. Most companies approached the FDA with projections and timelines pushing off the advent of the technology for a decade, but Dr. Reilly and Dr. Martha logically discussed what biotechnology knows today, and how to leverage that information for improved public health. The elements of pharmacogenetics are quite obvious. To predict drug response for a disease, three elements need to converge: first, a disease has to exist, second, a drug and its corresponding pharmacology must be understood, and finally, common gene variations associated with a drug target must exist and be understood. Five years ago, this convergence may have seemed a distant prospect because information about genes was scarce; however, the Human Genome Project has remedied this situation. Data about genes is plentiful, which is mirrored in drug development programs. Furthermore, technologies that can accurately and rapidly analyze genes are standard in many labs, and very affordable. There are very few reasons left not to codevelop genetic drug response profiles and pharmaceuticals; few medical reasons, that is.

While the FDA deliberates as to if and when it will require pharmacogenetics analyses in clinical trials and prescribing, there are other areas of concern in which Interleukin Genetics addresses the ethical issues surrounding biotechnology. Intellectual property, for instance, plays a critical role in most of biotechnology; however, the criticism surrounding patenting genes makes it difficult for any company to balance both social and corporate interests. Without their intellectual property, biotechnology companies have very little leverage in the marketplace. Although many firms bank on analytic technologies (machinery for analyzing DNA, for instance) or pharmaceuticals, other groups have to patent genes for their livelihood. While some companies avoid the questions concerning ownership of genes, others confront it head on. A group like Celera, which began its operations based on patenting as much of the human genome as possible, defends its position by claiming that most critics misunderstand the intricacy behind research and industry. Craig Venter, Celera's founder and spokesman, clarifies this belief in a recent statement: "There's a lot of people out there that want to make hay about anything we're doing. . . . It's scare-mongering, and it's doing harm to the American public's ability to understand this complex field."[29] To an extent,

---

[29]Gillis, Justin. Gene Researcher Draws Fire on Filings, Venter Defends Patent Requests. *The Washington Post,* October 26, 1999.

Venter is correct. Patents do not restrict academic research, just commercial applications; this distinction is fine, but important.

Despite the protection of patents, some firms do cling to their intellectual property without releasing information to potential collaborators. Although patents and the data used to win them exist in the public domain, those who own them usually have a substantial head start on related research. At Interleukin Genetics, the majority of the company's patents deal with three genes in the IL-1 gene cluster. Having covered most diagnostic and predictive applications in their filings, Interleukin has also amassed a substantial database of information around these genes, which is very valuable to the vast number of researchers focused on cytokine biology. While many groups might fear that sharing information and collaborating with an academic researcher would threaten their business objectives, Interleukin Genetics has taken the other route. Fostering research with many different universities and encouraging the investigation of the IL-1 gene cluster, Interleukin believes that its portfolio's value will only increase with the free exchange of ideas. Furthermore, should new intellectual property arise from these collaborators, the company enjoys an excellent strategic position. Because of a strong relationship based on trust and knowledge sharing, Interleukin Genetics has more than once been granted exclusive licenses to technology in which an interest exists, but the necessary resources to pursue it do not. Furthermore, provisional patent filings—patents filed to protect a preliminary idea where the company has either little scientific evidence or little cash to pursue it—may be validated through one of these collaborations. Through the free exchange of ideas, Interleukin Genetics has been able to develop a strong intellectual property portfolio and relationships with respected academic collaborators who represent an R&D department beyond the funding capability of the company.

Because the company focuses on a particular set of genes rather than broadly throughout the genome, Interleukin Genetics approaches its scientific goals differently than other firms investigating pharmacogenetics. Companies such as Variagenics and Genaissance try to identify as many SNPs as possible and associate them with disease. Indeed, the research efforts of the latter have resulted in a landmark paper that is forcing the biotech community to reevaluate its approach to polymorphic analysis. In this paper,

Genaissance analyzed over 300 genes, discovering that the human population's genetic variations are substantially higher than previously thought. Furthermore, expected haplotype patterns are disparate with the data; thus, statistical analytic techniques require revision. According to the findings, "The processes underlying genomic evolution are obviously subject to varying levels of natural selection, which invalidates overly simplistic theoretical models. Additionally, complex or unknown patterns of human migration complicate the distribution and interpretation of genomic variation."[30]

Posing a new scientific obstacle, this revelation also challenges a company's ethical position. Because there seem to be a much higher frequency and combination of polymorphisms in genes related to both disease and drug response, careful consideration must be given to an investigatory strategy. As indicated in the discussion on the ethics of pharmacogenetics, the genetic differences in developing countries can become a barrier to healthcare delivery if drugs are developed pharmacogenetically. That is, if individuals representing populations from poorer nations are excluded from clinical trials, the number of therapies suited to these people may diminish. The only way around such a predicament would be to characterize as many SNPs in a gene as possible, for as many genetically distinct populations as possible. Luckily, this approach meets the demands of both scientific and ethical inquiry. In terms of scientific inquiry, the more comprehensive the data, the better the interpretation. Ethically, a comprehensive database would be inclusive, rather than exclusive, targeting all individuals, as opposed to just a few.

Realizing this fact, Interleukin Genetics entered into a collaboration with Genome Therapeutics Corporation (GTC). According to this agreement, GTC would identify as many SNPs as possible in as diverse a population as possible. With this strategy, the IL-1 gene cluster will be "sequenced to the highest level of completion possible for SNP detection."[31] If this precision were limited to a small population, it would seem that only half of an important scientific question would have been asked. Furthermore, focusing on just part of that question may make sense from a market standpoint.

---

[30]Stephens, J. C. et al. Haplotype Variation and Linkage Disequilibrium in 313 Human Genes. *Science, 293,* 489–493, July 20, 2001.
[31]*http://www.ilgenetics.com/interleukin.htm.*

That is, it may seem worthwhile to develop therapies for just those populations who have the resources to pay for them; however, ethically, this approach is clearly suspect.

To Interleukin Genetics, helping the patient and collaborating freely
opens the door to strong business opportunities. Serving itself through serving others corresponds to one of the group's business objectives, described by
company co-founder Ken Kornman as "enlightened self-interest." As Dr.
Kornman further explains,

> Interleukin [Genetics] wants as complete a database as possible. We
> have worked with other companies on projects in the past, and we
> were surprised to find that they have very little ethnic diversity in their
> sample population, and it didn't make sense to us. We want to develop
> treatments for as many people as possible, and we want to answer im-
> portant scientific questions to get there. You can't know the answer to
> biological questions without asking them from as broad a standpoint
> as possible. With this strategy, we can meet our corporate goals, which
> in my eyes, means improving the world's general quality of health.
> While a lot of companies out there want to extend life by tinkering
> with things that have unknown implications, I'm not sure if that is a
> responsible route. But, if lifespan doesn't change through our work,
> but our efforts substantially improve the quality of life over that time,
> I'll be very happy.

VI. Recommendations

As mentioned earlier, pharmacogenetics and DNA data banking have quite
a bit in common. Because both depend on sample acquisition and clinical
trials to reveal genetic information about individuals and disease, many
ethical issues overlap. In particular, privacy and confidentiality are critical
components to both technologies, which means that the ethical recommendations outlined in the previous chapter are equally applicable to pharmacogenetics. Ranging from informed consent to data protection, reviewing the previous chapter's recommendations is *critical* to, not suggested for,

addressing the ethical issues surrounding pharmacogenetics. The reverse also applies for some of the recommendations in this chapter; data banking companies should thoughtfully consider the points presented in this section.

Any group positioned to reveal genetic information to an individual should make it a high priority to extract and deliver information through a counselor, not just a physician. Because genetic information is difficult to interpret, complicated, nuanced, and potentially frightening, a responsible group will do more than just relay the results, it will explore options with the patient to aid in psychological adjustments. As Barbara Bowles Biesecker, Co-Director of the Genetic Counseling Research and Training Program of the National Human Gnome Research Institute explains, "Genetic counseling is a short term psychotherapeutic relationship that incorporates the transmission of genetic information."[32] There are two important components brought out in this statement: first, the psychological impact of genetic information, and second, the transmission of information. These ideas are intimately related; explaining the data in layman's terms helps resolve any psychological issues that may be inflicted by this alien form of knowledge. Furthermore, the results may be difficult for the doctor to relay to a layman; thus, one of the counselor's responsibilities would be to translate the information into language whereby the implications to the patient are made apparent.

Genetic counselors have unique skill sets that place them in an excellent position to do more than just relay information to the patient. Their expertise is applicable in other areas, such as public relations or working with investors. With substantial explanatory capabilities, and fairly objective viewpoints in the best cases, they present details to whomever needs them, and, according to Diane Baker, Director of the University of Michigan's Genetic Counseling Program, "they won't sensationalize the science or frighten anyone."[33] Objectivity would also help a company to be more responsive to its

[32]Biesecker, Barbara Bowles. Privacy in Genetic Counseling. In *Genetic Secrets: Protecting Privacy and Confidentiality in the Genetic Era*. Mark A. Rothstein (Ed.), New Haven: Yale University Press, 1997, p. 108.

[33]Kling, James. Genetic Counseling: The Human Side of Science. The Scientist. *http://www.the-scientist.com/yr1999/july/prof_990719.html*.

stakeholders, and counselors would certainly facilitate such practices. Furthermore, support at all levels, from the scientific to the psychological, goes a long way toward convincing potential consumers, be they patients, doctors, or HMOs, of the firm's responsible nature. To reassure all stakeholders is to solidify an expansive customer base.

Although genetic counselors have a lot to offer to both biotechnology companies and patients, their skills can enhance the scientific process as well. Because counselors engage the patient prior to analysis, important scientific information, for instance pedigrees, will be extracted. This introductory period offers an excellent way to explain informed consent, privacy issues, and the repercussions of research, as well as add to the firm's database of information about a disease. Adding this skill set to a company's operations does more than just address patient concerns; leveraging the relationship without compromising it can substantially help a firm meet its corporate goals.

Although a genetic testing company, be it pharmacogenetic or diagnostic, ought to ensure that patients receive counseling before and after testing, the firm has to do its own "counseling" of the public as well. Pharmacogenetics, like most new technologies, meets some resistance because opponents are ignorant of relevant facts. That is not to say that criticisms are not deserved, but some may be misplaced. In cases such as this, the responsibility lies with the firm to educate the public. Executives often wonder how they can educate the world about these new technologies when many people do not have the basic intellectual background to decipher the explanations offered. One may counter, and quite effectively, by asking how it is that companies somehow know the most effective way to market equally complex products. While it may not be as simple as commanding the marketing department to tutor the general populace on the details of pharmacogenetics, it is worthwhile to identify different classes of students, and gradually approach each one. For instance, in order of increasing difficulty, the following groups ought to be educated: scientists, other companies (pharmaceutical or biotech), doctors, insurance companies, HMOs, regulators, elected officials, and the public. Although the order and inclusiveness of this list may be questioned, a point still remains. When marketing, different groups are separated into segments characterized by their likelihood of buying a product and the unique strategy used to reach them. The criteria used for this segmentation often reveals the type of knowledge assumed to be possessed by that population and to what extent they ought to be educated to convince them of the need for a product. Marketing is obviously more complex than the brief lines dedicated to it here, but the lesson to be learned from this exercise is that companies already classify groups according to the same criteria that can be used to target educational initiatives. Scientists do not need to learn the science, but a firm that understands the ethical implications of the research can share that knowledge with the scientific community at conferences or other gatherings. With physicians, the subject of genetics ought not be assumed as part of their training. Not only would the company have to explain genetics; it would also have to share the same information taught to scientists. As more and more of these groups are educated, the type of information conveyed and the methods for doing so in each preceding case will help the process each step down the line. Regulators, for instance, should be told all of the advantages to using pharmacogenetics in clinical trials; however, the social risks involved should be explained, and so on and so forth, until the end user, the patient, is reached. When appropriate information is relayed to patients at an individual level by the doctor, who may have been instructed by the HMO, the company selling the product, or anyone in this chain, then the firm will know that its efforts were successful and that the patient was afforded more than cutting edge care; he or she also freely chose this route through informed choice. Marketing to the individual may be the prevailing trend in healthcare; however, to assume that it can be done without an educational component and by only targeting one category of consumer misses the many steps in the path toward delivering an acceptable standard of care.

Part of the educational process, which also has a marketing component, is addressing the rhetoric surrounding personalized medicine. Pharmacogenetic analysis is a genetic test or genotyping assay to the extent that genes are being analyzed to reveal information about the health status of an individual, yet describing it as such is somewhat of a misnomer. Over the past few years, the power of this type of information has penetrated society through debates and discussions that both warn of and herald its adoption. Because of issues ranging from "stealing" someone's genes to the damage and stigmatization incurred by revealing such information to the world, many are wary of genetic tests. Indeed, the potential harm is frightening, but the potential good is equally astonishing. Because the term "genetic testing" carries so much baggage with it, researchers working in the field of pharmacogenetics need to explain the distinction between genetic analysis for personalized medicine and genetic analysis for disease diagnosis. Many of the concerns facing genetic testing do not apply to personalized medicine because the related analysis does not indicate disease like a diagnostic genetic test would. Rather, pharmacogenetic testing reveals the best course of dealing with disease. The condition is already known, and the question left for pharmacogenetics to answer on behalf of the patient is: which drug would deliver the greatest benefit and fewest side effects based on the patient's unique genetic profile?

To meet this rhetorical challenge, terminology much be chosen to highlight the distinction. Genetic testing will obviously not suffice, and pharmacogenetic testing may work; however, another qualifier for "genetic" rather than "pharmaco" may be necessary. The great difficulty lies in working around the term "genetic," which means that standard genetic testing may require renaming. Perhaps companies should indicate that "genetic" tests are not an option. Instead, both "diagnostic" genetic testing and "therapeutic" genetic testing could solve this nomenclature dilemma. The former denotes disease testing, whereas the latter informs prescriptive measures. Granted, overhauling language is a difficult task, especially in the scientific field where the specialized vernacular is as important to the culture as the scientific method; however, this approach may be necessary to serve the patient in the way espoused by this technology's proponents. The greater challenge is convincing the field to do so for ethical reasons, as opposed to economic or scientific ones.

When developing these "pharmacogenetic tests," a number of nonscientific choices have to be made because business interests have to be met. Simply put, if corporate needs are not satisfied, then the company may not survive long enough to bring diagnostics and drugs to the public. Financial interests cannot be ignored; thus, these firms meticulously choose particular diseases as targets based on market size, treatment frequency, and other considerations. However, taking this reasoning too far must be avoided, as temptation may lead to two problems contrary to public health. The first is concerned with "orphan" drugs, and the second centers on the predictive power of a pharmacogenetic test.

Orphan drugs are treatments that correspond to rare diseases. Because of their infrequency, the cost of development is rarely offset by the sales of the product. At present, government agencies in both the United States and the European Union offer incentives to pharmaceutical companies to develop treatments for rare disorders through tax credits, allowing time-limited monopolies, and other initiatives.[34] However, when pharmaceutical companies begin developing a series of drugs for "popular" disorders, rather than single blockbuster drugs, research efforts may be spread over too many projects critical to a company's success to allow research for uncommon diseases. Two avenues must be pursued to prevent this. Pharmacogentics companies will have to devote resources to these types of drugs and convince policy makers, as bioethicists Mark A. Rothstein and Phyllis Griffin Epps argue, to "consider whether to expand the concepts underlying orphan drug policies to stimulate research into and the development of drugs for populations who, by virtue of their genetic makeup, face inequities in drug development efforts."[35] The burden should not lie on the shoulders of the government to make it attractive for the pharmaceutical industry to develop such treatments; the pharmaceutical industry should take the steps to reach a compromise with the powers that be to continue the development of such programs and therapies.

Genetic equity will be revisited below, but the second implication of favoring business over social interests requires discussion. Focusing treatment programs on "popular" diseases is a two-edged sword. On the one hand, some of these diseases already have blockbuster status, while on the other, not all patients taking such a drug are actually responding to it. Although it is safe to assume that corporations will develop other drugs to make up for

[34]Thamer, M. et al. A Cross-National Comparison of Orphan Drug Policies: Implications for the US Orphan Drug Act. *J. Hlth. Politics Policy Law, 23,* 265–290. 1998.
[35]Rothstein, M. A. and Epps, P. G. Ethical and Legal Implications of Pharmacogenomics. *Nature Reviews Genetics, 2,* 230, March, 2001.

the failure of the current option, that does not mean that pharmacogenetics will be utilized to the full benefit of the patient. The temptation in this scenario is to limit research to identifying side effects rather than response profiles, at least until a panel of drugs for the disease comes to market. For instance, a leading allergy medicine may cause side effects, and a manufacturer may perform the clinical trials to identify which patients ought to and ought not take the drug based on adverse reactions. However, until the same company develops another allergy medicine, it may not release any of its drug response information. If it did, a competitor with an effective alternative for those nonresponders may assume a substantial proportion of a previously closed market share. Using allergies as an example does not relay the magnitude of this strategy's offense. If the disease were something degenerative, like rheumatoid arthritis, the degree of harm becomes clearer. In a case such as this, where the medicine not only relieves symptoms but also slows the progression of the disease, withholding a pharmacogenetic response test not only perpetuates the pain, but may also aid disease progression. Ethically, companies need to recognize patients as people suffering from disease who place equal amounts of trust in and dependence on industry to help them. To betray that trust and take advantage of that dependence in favor of meeting an economic bottom line is to reify the reputation that large, multinational corporations behave as heartless engines of capital. Patients are stakeholders in pharmacogenetics, and to meet their needs, firms need to focus as many resources on developing and promoting drug response tests as they do on side effects testing.

Issues of genetic equity do not only center on disease. Because the power of genetic information places it in a realm affecting more than health matters, pharmacogenetics requires ethical interrogation in social and political terms, as well. As Rothstein and Epps point out, "Compared with traditionally designed human clinical trials, genotype-specific human clinical studies might be subject to equal or greater limitations in that the relatively short duration of the study, combined with narrower subject population and smaller size, would hinder the ability to identify rare or delayed adverse reactions or drug interactions."[36] Coupling this viewpoint with the

---

[36]Ibid., 228.

prevalence of genetic variations associated with ethnic diversity points towards another potential level of genetic inequity. Rothstein and Epps focus on poor side effect data for ethnic groups excluded from clinical trials, but continuing the argument reveals an even greater concern. Limiting the ethnic diversity of a clinical trial cohort may not only restrict the amount of data correlating drug side effects with the excluded group, it may also lead to developing drugs that respond to specific racial groups. Although the obvious discriminatory issues arise, previously unrecognized concerns emerge in the area of global healthcare rationing. Socioeconomic factors already place the best medical care in the wealthiest populations. This same reapportionment exists on a global scale; richer nations enjoy greater access to therapies, and poorer nations make due with less advanced solutions. Unfortunately, poorer nations are also populated by racial groups that have traditionally comprised the lower socioeconomic classes in developed countries. Although standard economic and political arguments rationalize this pattern, that does not excuse the fact that scientifically, this schism may increase if not addressed. Thankfully, pharmacogenetics is such a new field that this problem can be addressed before it actually increases the current gap in global healthcare. Pharmacogenetic research must take into consideration that ethnic variation may play an important role in drug response profiles and drug development. The risk is that a lot of time and money may be spent on some drugs that do not have variable response across geographically or ethnically diverse populations; however, if they do, not asking the question may result in more than just compromising public health. The possibility that science may help in solidifying already contentious global healthcare practices must be avoided. Furthermore, assuaging many racial groups' fears of biopiracy must also occur if the experimental material needed to reach conclusive results in developing therapies is to be obtained.

How to ethically deal with intellectual property issues is somewhat difficult to navigate because patents are vital to a company's success. Thus, not patenting a novel, functional gene can never be an option. Because patents are restrictive by nature, corporations should consider where it is that licensing deals and other royalty structures can be loosened in favor of social interests, as well as where it is that business interests demand very restrictive

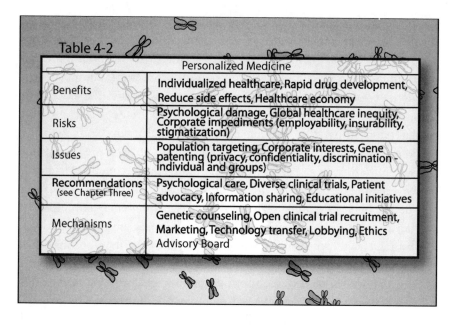

Table 4-2	Personalized Medicine		
Benefits	Individualized healthcare, Rapid drug development, Reduce side effects, Healthcare economy		
Risks	Psychological damage, Global healthcare inequity, Corporate impediments (employability, insurability, stigmatization)		
Issues	Population targeting, Corporate interests, Gene patenting (privacy, confidentiality, discrimination - individual and groups)		
Recommendations (see Chapter Three)	Psychological care, Diverse clinical trials, Patient advocacy, Information sharing, Educational initiatives		
Mechanisms	Genetic counseling, Open clinical trial recruitment, Marketing, Technology transfer, Lobbying, Ethics Advisory Board		

measures. For instance, some physician groups have issued statements advising that corporations approach genetic tests, pharmacogenetic and otherwise, differently than other diagnostics. More precisely, the American College of Medical Genetics and the Academy of Clinical Laboratory Physicians and Scientists believe that these procedures should be broadly licensed for reasonable fees, so long as the testing lab meets certain standards.[37] In general, sharing information that will lead to better research initiatives should not prove too challenging. Indeed, the very point of a patent is to protect the novelty of a group's or individual's efforts and to ensure that the work is rewarded. Collaborating freely may lead to ownership of more intellectual property or even a stronger patent position.

Along with its tremendous potential, personalized medicine carries both old and new ethical concerns. Although it has yet to reach its promise, it will

---

[37] American College of Medical Genetics, Position Statement on Gene Patents and Accessibility of Gene Testing. *http://www.faseb.org/genetics/acmg/pol-34.htm*. Academy of Clinical Laboratory Physicians and Scientists. ACLPS resolution: Exclusive licenses for diagnostic tests approved by ACLPS Executive Council 06/03/00.

some day force a shift in healthcare by allowing clinicians to end their one size fits all approach to medicine. Using genetic information as a basis of healthcare will leverage what was once a confounder of such care. If companies focused on this technology can perform their work ethically, then medicine stands to become what the life science industry has always promised it would be: the best care for all.

5

STEM CELLS

... let abortion be procured before sense and life have begun; what may or may not be lawfully done in these cases depends on the question of life and sensation.

—Aristotle, *The Politics*

I. Executive Summary

Despite its incredible accomplishments, there are still substantial limitations as to what medical science can achieve. In their most challenging moments, patients and their doctors are faced with failed organs and other dead tissue that cannot be regenerated, as well as mysteries about development that make treating embryos incredibly difficult. Despite their seemingly disparate bases, these problems have a very common, potential solution. Embryonic stem cells offer both a means of regenerating tissue and a window into the complexity of development; however, the history of medicine has clearly revealed that research surrounding embryos is socially volatile, and this technology is no exception.

The Science

Embryonic stem cells are, as their name suggests, derived from embryos, and their unique characteristics allow them to be cultured indefinitely. Furthermore, their early place in the developmental cycle endows them with the potential to become any cell in the human body, thus they can be used to re-generate dead tissue. These same characteristics make them attractive candidates for developmental biologists looking to unlock the mysteries of development.

The Benefits

Because of their regenerative capabilities, these cells may hold the key to curing conditions previously thought beyond the immediate capabilities of modern biology. For instance, Alzheimer's disease, diabetes, Parkinson's disease, and spinal cord injuries are but a few of the many afflictions that this technology might cure and/or prevent. Furthermore, these cells can be used to develop safer drugs and offer insight into reproductive processes.

The Issues

There may seem to be many issues that relate to this technology, but they are all subsumed by the major ethical challenge posed by it: What is the moral status of an embryo? Although many disagree on this point, it is because of a disagreement over the definition of life, predicated on an agreement over life's sanctity. The challenge lies in leveraging areas of agreement to address the concerns of those that disagree.

The Industry

Two companies are used to explain how industry has and can deal with the ethical issues surrounding stem cell research. Both take very different stances, and they show that corporations have numerous options in front of them, but strategies must be chosen carefully and rationally. In the case of Geron, the company believes in the importance of this research, but at the same time they understand that the ethical issues are a very important part of this technology's development. This fact has prompted Geron to inform the debate by making itself accessible to concerned groups, all the while being careful not to influence the debate. The other firm profiled, Advanced Cell Technologies, believes so strongly in the technology that it sees its moral obligation as one of educating the public. Afraid that the public will not adequately engage this discussion, Advanced Cell Technology tries to keep the issues in the public sphere so that the issues can be thoroughly addressed by those most strongly affected by them: society.

Recommendations

Although there is no clear way to prove the morality of this research, whether one believes it good or bad, there are steps that can be taken to obviate the objections held by many. For instance, "appropriate" modes of deriving these cells are discussed, as well as how consent procedures should be maintained to inform those involved of the issues. To maintain a pattern of responsibility, this chapter also suggests that an ethics advisory board be appointed to advise on research and corporate decisions regarding the technology.

If you could end all human suffering, would you do it? What about in the context of disease; if you knew that you could vanquish all illness, how quickly would you make the decision to do so? Few people would hesitate if presented with an opportunity to end the misery associated with disease; however, if the price of such a boon was the life of one child, and the decision was solely in your hands, would you make that sacrifice and justify it in the name of the greater good? This mental exercise is one of the most basic teaching tools in moral philosophy. Many undergraduates are presented with this dilemma, only to forget it after the semester ends, asking a different question as they leave the classroom: "When will I ever need to use what I learned about addressing this problem?" For many people involved with stem cell research, this question pervades every detail of their investigations.

Embryonic stem cells offer one of the most promising therapeutics in the history of science; however, because this tissue is derived from embryos at the expense of the fetus' survival, the technology is mired in the ethical imbroglios surrounding the abortion debate. For those who consider themselves a part of the pro-life movement, the question is exactly as stated above; the price of cures versus the price of lives. The unwavering answer of the pro-life movement is: unborn life should almost always prevail. However, the opposite camp, the pro-choice movement, believe that there is no moral dilemma because a child's life is not at stake; rather, the opposition is wasting their breath on a nonsentient cell mass.

The promise of this technology has placed it in a position in which advancing it is almost a moral imperative for the companies working on it. These remarkable cells can be coaxed into becoming any tissue in the human body, making degenerative diseases or any affliction that results in cell death curable when the technology is perfected. As Michael Werner, head of legal and ethical issues at the Biotechnology Industry Organization (BIO) has stated:

> The discovery of pluripotent stem cells may be the single most important scientific and medical breakthrough of the past decade. This work has the potential to impact the lives of millions of Americans suffering from many of humanity's most devastating illnesses, including Alzheimer's and Parkinson's disease, diabetes, heart disease, cancer, and spinal cord injury. There is great hope that pluripotent stem cell research will result in new treatments and cures for many of these diseases and disabilities.[1]

Industry finds its justification in statements such as this, as well as the arguments used by the pro-choice movement, but these defensive mechanisms will only go so far in light of social and ethical concerns.

At issue here is the moral worth of the human embryo. One side believes that an embryo is equivalent to a human being and the other claims that it is nothing more than a mass of cells. Biotechnology, however, is in the habit of altering our very conceptions of nature, as well as our definitions of life, and although the abortion debate has been framed by a strict dichotomy, this new research is upsetting the balance. There are types of research that are being performed on human embryos that are now finding support by people

---

[1] *http://www.bio.org.*

who classify themselves as pro-life, while some pro-choice activists oppose certain, related investigations. Ideologies are being challenged, and the black and white worlds separated by embryos are interrupted by shades of gray, as evidenced by a statement made by Republican Senator Gordon H. Smith. An avowed pro-life supporter, Smith dissolves party lines because he sees a clear rationale for encouraging stem cell research: "Part of my pro-life ethic is to make life better for the living."[2] Interestingly, he has since been joined by a number of his Republican colleagues who hope to convince President George W. Bush to lift the ban on embryo research with regard to funding stem cell experimentation. However, ban or no ban, the issue will remain a substantial one.

Partially converted Republicans should not be viewed as a victory for abortion supporters; there are also conversions occurring in the other direction. A woman's right to choose and performing research on an embryo are related, but not equal concerns. The latter may be used as a justification for the former, but the elements that surround experimentation on embryos is unsettling to some supporters when faced with the reality of the process. Furthermore, when technology can remove the woman's body from this equation, as in the case of in vitro fertilized embryos (IVFs), the woman's right to choose is no longer the prevailing ideology. When displaced from this framework, interpreting life's value in the context of creating embryos for experimentation challenges ideas that are still being worked out in our culture. For instance, few people read Aldous Huxley's *Brave New World* as a manual for conduct; most recognize it as the cautionary tale that it is. The distressing imagery that Huxley conjures in his opening pages speaks to the concerns that convince pro-choice supporters that this research is blurring the lines between themselves and the opposition. To have enough cells for clinical and investigatory purposes, some stem cell research supporters contend that embryos need to be created for this sole purpose, which is a concept that is not always supported or contested according to the usual party lines. In the Bokonovsky process described by Huxley, the literary portrayal he meant to warn of is, technically, almost the exact process under contemplation today:

---

[2]Stolberg, Cheryl Gay. Stem Cell Research Advocates in Limbo. *New York Times,* January 20, 2001.

One egg, one adult-normality. But a bokanovskified egg will bud, pro-liferate, will divide. From eight to ninety-six buds, and every bud will grow into a perfectly formed embryo, and every embryo into a full sized adult. Making ninety-six human beings grow where only one grew before. Progress.

Progress in the case of stem cells does not lead to human beings; the process will never get as far as *Brave New World*'s Hatchery. These embryos are des-tined for scientific advances, producing cell lines that can become any tissue in the human body, increasing corporate valuation, aiding humanity, and "death." As Huxley said, "Progress." The question now is not just weighing the life of a child against a cure for disease; rather, as corporations spearhead this research, the dilemma that needs to be addressed centers on balancing corporate beliefs, progress, and social responsibility: how can corporations continue their work in a manner in which social harm is not the byproduct of progress?

II. The Science

As with many biological discoveries, the scientific promise of stem cells be-gan in early experiments with mice when, in 1981, researchers isolated the first stem cells from mouse embryos.[3] These embryonic stem cells (ES cells) exhibited some startling characteristics, which greatly appealed to the imagination of the scientific community. Able to grow rapidly for an ex-tended period of time, these cells could do much more than proliferate readily; they could also differentiate into any type of mouse tissue if placed in an accommodating environment.[4] Furthermore, these cells could be

---

[3]Evans, M. J. and Kaufman, M. H. Establishment in Culture of Pluripotential Cells from Mouse Embryos. *Nature, 292,* 154–156. Martin G. R. Isolation of a Pluripotent Cell Line from Early Mouse Embryos Cultured in Medium Conditioned by Teratocarcinoma Stem Cells. *PNAS, 78,* 7634–7638, 1981.

[4]Robertson, E. and Bradley, A. Production of Permanent Cell Lines from Early Embryos and their Use in Studying Development Problems. In *Experimental Approaches to Mammalian Embryonic Development,* J. Rossant and R. A. Pedersen (Eds.), New York: Cambridge Univer-sity Press, 1986, 475–508.

transferred to a mouse embryo and indistinguishably become incorporated into the embryo's normal developmental pathway. These findings would lead to the recognition that stem cells could be coaxed into developing into cells that are identical to any mouse's organ's cells. Realizing that these cells held the immense potential to replace damaged tissue, it would take scientists over a decade before similar experiments would be successful in human embryos.

In 1994, Bongso and colleagues isolated the first stem cells from a human blastocyst, and four years later, James Thomson's group at the University of Wisconsin would successfully culture the cells, providing enough tissue for further experimentation.[5] Before delving into how these cells are used, however, a brief explanation of human development and the derivation process is necessary. After fertilization, the human egg begins a stepwise developmental process and, after about nine months, differentiates into a child. The first cell divides into two cells, and those two cells divide, continuing the process of creating identical cells until their number reaches sixteen. Previously a zygote, this cell mass is known as a morula, which leaves the mother's fallopian tube and enters the uterine cavity. Division continues until the cells number in the hundreds and form two different types of tissue. An exterior mass surrounds an internal cell cluster known as a blastocele; together these cell groupings are called a blastocyst. This stage is too early in development for any of the cells to have been "programmed" to become a specific type of human cell, but they are also too far along in development for any single cell to become an entire human being. Each cell, then, has the ability to become any cell in the human body, and eventually a specific set of signals will differentiate the cells, that is, tell them what part of the body they are destined to become. The blastocyst stage is reached within six to seven days after conception, and the cluster will not progress to the next stage, when "programming" begins, for about another week. After that second week following fertilization, the cells lose the potentiality that make them attractive as stem cells; at this point they begin the first round of specialization known as gastrulation, meaning they no longer have the ability to become any cell in the human body (Figure 5-1).

---

[5]Bongso, A. et al. Isolation and Culture of Inner Cell Mass Cells from Human Blastocyts. *Human Reproduction, 9* (11), 2110–2117, 1994. Thomson, J. A. et al. Embryonic Stem Cell Lines Derived from Human Blastocysts. *Science, 282,* 1145–1147, 1998.

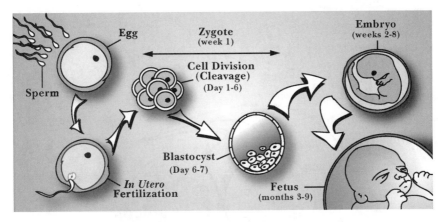

Figure 5-1   Development of an embryo.

Stem cells are derived from the blastocyst stage of development. Although the outer cells that contain the blastocele act as protection and help the morula implant itself in the uterus, the inner cell mass contains the stem cells. Stem cells are defined by the National Institues of Health as "cells that have the ability to divide for indefinite periods in culture and to give rise to specialized cells."[6] Put in technical terms, these cells are described as pluripotent, as opposed to totipotent. Primitive cells in this context are understood as one of these two types: either pluripotent or totipotent. Preceding pluripotent cells in the developmental process, a single totipotent cell has the ability to mature into an individual organism. Pluripotent cells are further along the developmental pathway, and although an individual cell does not have the potential to develop into a single organism, it can differentiate and proliferate to form any type of cell or tissue in an organism. Furthermore, these stem cells not only find and repair damaged tissue; they can also replace dead cells with younger versions, meaning that the replacements are the equivalent of newborn, unaged cells with a much longer future than adult cells.

These pluripotent cells can also be derived from embryos further along the developmental pathway, but only in the form of germ cells. Germ cells will eventually become gametes (sperm or ova), and maintain their pluripotentency for a much longer period than the other cells in the devel-

[6]Stem Cells: A Primer. *http://www.nih.gov/news/stemcell/primer.htm.*

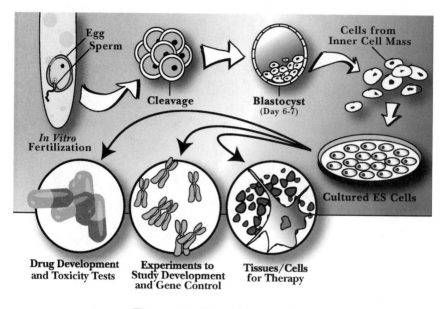

Figure 5-2   Harvesting stem cells.

oping embryo. Because these primordial germ cells (PGCs), as the are also called, can be isolated and used as stem cells well past the blastocyst stage (which lasts up to 14 days after fertilization), the PGCs can be harvested from aborted fetuses instead of from an early stage morula. At about the same time that Thomson isolated and cultured blastocyst-derived stem cells, John Gearhart's lab at Johns Hopkins University did the same with PGCs.[7]

After isolating these cells, they are cultured to grow into enough copies as may be experimentally or clinically useful. Here lies another advantage of stem cells, the ability to proliferate in culture "indefinitely." Although there are limits as to how long normal divisions will continue, these cells multiply at a rate unlike any other human cell type. Put together, all of the abilities of stem cells (Figure 5-2) offer tremendous therapeutic potential, as described by the National Bioethics Advisory Commission (NBAC):

[7]Gearhart, J. D. New Potential for Human Embryonic Stem Cells. *Science, 282,* 1061–1062, 1998.

Because stem cells are able to proliferate and renew themselves over the lifetime of the organism—while at the same time retaining all of their multilineage potential—scientists have long recognized that such cells could be used to generate a large number of specialized cells or tissue through amplification, a possibility that could allow the generation of new cells that would treat injury or disease.[8]

Despite the tremendous potential of embryonic stem cells, there do exist adult stem cells (AS cells or ASCs). For instance, scientists have known of hematopoietic, skin epithelium, intestinal epithelium, and neuronal stem cells for quite some time. Each of these types of cells is a source of tissue that human bodies regularly need to regenerate. Hematopoietic cells give rise to all types of blood cells; the other types do the same in their respective roles, promulgating skin cells, intestinal cells, and neural cells. However, ASCs have limited utility because they are specialized toward developing into specific sets of cells rather than the diverse range that embryonic stem cells can produce.

Research is under way to determine whether or not AS cells can expand their applicability. A number of investigations have shown tremendous promise for AS cells in animal and human models alike, but the results have yet to reach the efficacy of the ES cells. Worldwide, scientists are working hard to discover a means to avoid embryo research and bring the less offensive technology surrounding AS to the advanced level of ES cells. After performing experiments on sheep, Alan W. Flake, director of the Children's Institute of Surgical Science at The Children's Hospital in Philadelphia remarked, "The transplanted cells migrated to different parts of the sheep's body and differentiated into types of tissue at each site."[9] The cells remained active for over a year, and are suspected to be involved in healing wounds.[10] Within mice, Italian scientists have found that neural stem cells can be

---

[8]National Bioethics Advisory Commission. *Ethical Issues in Human Stem Cell Research,* volume I. Rockville, MD: Author, p. 7, 1999.

[9]News Analysis: Technology. Awaiting the Miracles of Stem-Cell Research. *Business Week,* November 29, 2000. *http://www.businessweek.com/bwdaily/dnflash/nov2000/nf20001129_ 858. htm.*

[10]Liechty, K. W. et al. Human Mesenchymal Stem Cells Engraft and Demonstrate Site-Specific Differentiation ater In Utero Transplantation in Sheep. *Nature Medicine, 6* (11), 1282–1286, November, 2000.

placed within muscle tissue and coaxed into acting as a part of the tissue if they are in close proximity to mature muscle cells.[11] Even more promising are the results of a University of South Florida group, whose experiments on stem cells found in bone marrow show that these ASCs can become immature nerve cells if placed in the brain of a patient with Parkinson's disease.[12]

Until (if at all) the AS cells are as generally capable as ES cells, ES cell research will continue to the point where the technology has clinical value. There are other uses for these cells, which will be discussed in the following section, but the technological hurdles involved with tissue replacement have an important scientific component that deserves attention in this part of the discussion. At its current stage, the research is focusing on how to derive, culture, proliferate, and deliver stem cells to a patient in need. These major obstacles have been overcome, although delivery still requires a great deal of testing, followed by the clinical trials that are involved in finalizing any type of human therapeutic. AS cells will face similar obstacles, but the substantial head start on these applications possessed by ES cells will hopefully obviate any major technical impediments that AS cells will face if they finally reach ESCs' advanced stage of development. Currently, one of these concerns is unlimited cell proliferation, which is an attractive characteristic in creating enough cells for a procedure, but is separated by a very fine line from cancer if it proceeds uncontrollably. A number of mechanisms can be used to prevent catastrophe, ranging from engineering the cell divisions to be temperature sensitive to discovering the natural process that halts stem cell development after differentiation. John Gearhart and his team of researchers followed up their 1998 breakthrough two years later with the potential answer to concerns over tumor formation. In December 2000, Gearhart's group reported that stem cells could be cultured to form embryoid bodies (EBs); from these cell clusters, EB cells are isolated and cultured in separate nutrient environments, forming embryoid body derived (EBD) cells.[13] According to Gearhart, "We can't take the embryonic (stem) cells that we have

---

[11]Galli, R. et al. Skeletal Myogenic Potential of Human and Mouse Neural Stem Cells. *Nature Neuroscience, 3* (10), 986–991, October, 2000.

[12]Sanchez-Ramos et al. Adult Bone Marrow Stromal Cells Differentiate into Neural Cells In Vitro. *Experimental Neurology, 164* (2), 247–256, August 1, 2000.

[13]Shamblott et al. Human Embryonic Germ Cell Derivatives Express a Broad Range of Developmentally Distinct Markers and Proliferate Extensively *In Vitro. PNAS, 98*(1), 113–118, December 26, 2000.

and transplant them into anything. . . . [They will] just as likely form a tumor as they will differentiate into some sort of tissue," but EBD cells are further along the developmental pathway and are specialized beyond uncontrolled proliferation; as Gearhart explains, "Once a cell commits, it won't become a tumor."[14]

Another problem that requires attention is the possibility of rejecting stem cells. Immune reactions are a standard concern for any type of transplantation, and there is the potential for a situation similar to the common graft versus host disease scenarios that plague organ transplantation. When an invader (as transplanted cells may be recognized) enters the human body, the body treats the transplant as a foreign disease, and a severe immune response attacks the trespasser. In organ transplantation procedures, the problem is overcome by two methods; first, as compatible a donor as possible is found via tissue typing, and, second, a battery of immunosuppressant drugs are prescribed to weaken the immune response. Impairing the immune system has obvious risks, and neither measure guarantees that the transplant will be accepted; luckily, stem cells can be controlled to avoid or mitigate these scenarios. Using a procedure known as somatic cell nuclear transfer (SCNT), commonly referred to as cloning, the stem cells can be transformed into seemingly identical cells to the host's. In this procedure, the stem cells are incised and then have their nucleus removed by very precise cutting and suction tools similar to syringes and pipettes. After the nucleus of the cell is removed, the nucleus from any of the donor's somatic cells are fused with the stem cell, giving the stem cells the host's genetic imprint. Culturing the cells creates enough of the new variety to be used clinically, and the ES cells are now genetically identical to the recipient of the transplant, protecting the host from an immune response and rejection (Figure 5-3).

III. The Benefits

In a statement made before the United States Senate Appropriations Committee, then NIH Director Harold Varmus explained the areas of research in which stem cells would open entirely new avenues of research:

---

[14]Reuters Health Information. Scientists Clear One Hurdle for Using Stem Cells. January 3, 2001. *http://www.heartinfo.com/reuters2001/010103elin006.htm.*

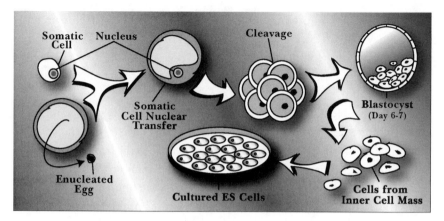

Figure 5-3   Somatic cell nuclear transfer.

Let me mention just three potential applications of human pluripotent stem cells. The first is research focused on how stem cells differentiate into specific types of cells. The goal is to identify the genetic and environmental signals that direct the specialization of a stem cell to develop into specific cell types. Studying normal cell and tissue development will provide an understanding of abnormal growth and development which, in turn, could lead to the discovery of new ways to prevent and treat birth defects and even cancer.

A second and more practical application of research using these cells is in pharmaceutical development. Use of human pluripotent stem cells could allow researchers to study the beneficial and toxic effects of candidate drugs in many different cell types and potentially reduce the numbers of animal studies and human clinical trials required for drug development.

The third and most obvious potential application of these human pluripotent stem cells is to direct the specialization of the cells into cells and tissues that could be transplanted into patients for the purpose of repairing injury and pathological processes.[15]

---

[15]Varmus, Harold. Statement Before the Senate Appropriations Subcommittee on Labor, Health and Human Services, Education and Related Agencies. January 26, 1999. *http:// www.nih.gov/news/stemcell/statement.htm.*

These remarks were made following the Department of Health and Human Services' recommendation to allow federal funding for stem cell research, despite their derivation from embryo tissue. Excited by the tremendous opportunity that this technology offered, Varmus hoped to enlighten Congress as to how the research will revolutionize medicine. An accomplished scientist and orator, not to mention Nobel laureate, Varmus very concisely mapped out the areas of research poised for innovation due to these remarkable cells.

Over the past few decades, developmental biology has attempted to understand the process of embryo development, but the inability to simulate the development of a human embryo coupled with the legal and ethical restraints on this type of experimentation has made it very difficult to peer into the early stages of the process. Since many of the questions surrounding this field focus on the specialization of cells, stem cells offer a means of unearthing the complex mechanisms that direct each cell to assume its ultimate physiological characteristics. As James A. Thomson and colleagues state in their original *Science* paper on the derivation of stem cells,

> Human ES cells should offer insights into developmental events that cannot be studied directly in the intact human embryo but that have important consequences in clinical areas, including birth defects, infertility, and pregnancy loss. Particularly in the early postimplantation period, knowledge of normal human development is largely restricted to the description of a limited number of sectioned embryos and to analogies drawn from the experimental embryology of other species.[16]

Cancer therapy is another area that may benefit from stem cell research, particularly chemotherapy. In chemotherapy, a drug or series of drugs are used to destroy the tumor by disrupting the cancer cells' reproductive process; however, these drugs have a tendency to affect normal cells as well, resulting in a number of side effects. Among these side effects is the depression of blood cell levels in the body, which may impair the body's own immune response to the cancer. Currently, bone marrow is transplanted into

---

[16]Thomson, J. A. et al. Embryonic Stem Cell Lines Derived from Human Blastocysts. *Science, 282,* 1145–1147, November 6, 1998.

the patient to replenish these cells, but the bone marrow stem cells do not display great efficacy in the elimination of cancerous cells. Injecting ES cells, however, may provide a better method of rejuvenating blood cells, thus aiding the body in its natural fight against cancer.

Aside from fighting disease, stem cell research will also aid in understanding the nature of disease, while broadening the mechanisms used to treat it. Toxicity studies are requisite in drug development; however, the difficulties facing such studies center on a dearth of in vitro analytic methods. Because the liver is the most likely organ to be damaged by experimental pharmaceuticals, drug toxicity testing is performed on cultured liver cells from cadavers long before the treatment is tested in living humans. VistaGen's CEO, Ralph Snodgrass, explains that the liver is "the major clearing house for drugs in the body," and the company has become very excited about the fact that cultured ES cells exhibit, if not match, many of the same characteristics of human liver cells.[17] Furthermore, the ES cells' ability to proliferate in culture makes them far superior to in vitro liver cells since the latter, in culture, tend to rapidly lose the ability to break down chemicals into their toxic byproducts. As a result of this methodology, drugs will come to market through safer and faster clinical trials.

In terms of clinical utility, stem cell science shows the most promise in repairing damaged tissue. Already, stem cells are being considered as treatments for degenerative disorders. Within the nervous system alone, diseases such as Parkinson's, Alzheimer's, and Huntington's typify the types of ailments that may be slowed or corrected with stem cell therapies. For instance, Parkinson's disease is characterized by the loss of nerve cells that generate dopamine; however, by introducing fetal tissue into the unhealthy region of the brain, scientists have found that they can slow or halt the disease.[18] Similar experiments and testing may lead to regenerative therapies that replace heart cells damaged by heart disease, or the missing islet cells in

---

[17]Cohen, Phillip. Toxic test: The Mother of All Cells Makes an Excellent Lab Guinea Pig. *New Scientist,* June 5, 1999. *http://www.newscientist.com/nsplus/insight/clone/stem/toxictest. html.*

[18]Freed, C. R. et al. Double-Blind Controlled Trial of Human Embryonic Cell Transplants in Advanced Parkinson's Disease: Study Design, Surgical Strategy, Patient Demographics and Pathological Outcomes. Presentation to the American Academy of Neurology. April 21, 1999.

the pancreas that are responsible for type I diabetes, which has already been accomplished in mice. Furthermore, coupling the technology with SCNT, a set of cultured stem cells can become "universal donor cells," that is, cells that can become any missing tissue in *any* human body. The therapeutic value of stem cells is limited only by the types of diseases and injuries that damage human cells.

IV. The Issues

Of the many technologies emerging from biotechnology, stem cell research is easily one of the most controversial. Despite the tremendous benefit to life that the technology offers, ES cells cannot escape the abortion debate because their derivation requires the "death" of the source embryo. Although the pro-life/pro-choice dispute may seem less relevant due to the scientific value of the research, both sides are well aware of the place that this technology takes in their respective camps. Pro-lifers see the technology as another violation of their beliefs, and, more damagingly, a potential excuse for the slaughtering of innocent life. Pro-choice activists believe that ES cells are a biological phenomenon that ought to be studied for the good of humankind, similarly to any nonhuman cell mass. Complicating matters in this case is the different sources of these cells, ranging from manmade organisms to aborted fetuses.

The abortion debate has received much attention from philosophers, religious representatives, scientists, political scientists, and many other intellectual groups. Although it is not the intention of this book to go into the fine points of each side, a brief survey of views will help in the following discussion. Despite the many people who have engaged this topic, it is still one of the most emotionally charged topics related to bioethics; consequently, many of the arguments are built upon preconceived attitudes. That is, very few people come to this topic without their mind already made up, and they will often let their emotions get the better of their intellectual training and use their knowledge to support their cause, rather than using their knowledge to critically examine the questions. This is not at all a new phenomenon in the world of ideas—Scottish philosopher David Hume described this disposition centuries ago when he said that "Reason is, and ought only to be, a slave to the passions."[19] Hume, however, had it in mind that emotions set the human agenda, and it is the role of reason to realize that agenda. The difficulty in this case is that neither side can be reasonable enough to empathize with the other's position, and emotions from two very powerful factions are in perfect conflict. Furthermore, the religious motivation of many of the pro-life activists places the foundation of their arguments in "faith" and "the word of God," which are absolute, and beyond earthly, human reason in the eyes of supporters. It is clearly a case in which the only harmony lies in the agreement to disagree.

Perhaps it is a bit hasty to characterize both positions as completely, ideologically dissonant. It is a tremendous irony that both groups greatly value the concept of human life; furthermore, moral worth is ascribed to all humans, thus they agree that it is immoral to destroy human life. It is the definition of human life that is confusing. As is well known, conception constitutes human life in the eyes of the pro-life movement, whereas the pro-choice activists believe the humanness requires more than the fusing of sperm and ova. To understand this debate, each group's reasons for their passion require investigation.

Members of the pro-life faction, aligned with political conservatives in the United States, believe that the beginning of life occurs at conception. Since life begins at that moment, special effort must be made by society to protect that life. Religious supporters of the pro-life movement see this miracle as more than a biological phenomenon, but the moment that God imparts a soul to a new human being, placing a piece of divinity within humanity. Respect for this new person, conservatives will argue, demands that its life be protected, not frustrated by science, governments, or any individual, be it a scientist, a Congressman, or the mother, whose beliefs conflict with this sanctity.

While the religious component of the pro-life movement has much to say, it must regrettably be removed from the discussion from here on in. There are too many religions that have a stance on this debate, and to give primacy to one or another would show a favoritism that has no place in this text. It may

---

[19]Hume, David. *Treatise of Human Nature*, 2nd ed. Selby-Bigge and Nidditch, P. H. (Eds.), Oxford: Clarendon Press, 1978, 415.

seem that some groups are more involved than others, but the truth is that religious groups only become passionately involved with the debate if they oppose abortion; no religion crusades for its approval. Admittedly, there is a weakness to this argument in that there is strong religious opposition to this practice, but the better reason for excluding this viewpoint from the text is because there is no way to rationally engage this argument. To discuss the presence of a soul and the divine spark that affords moral worth to humanity is to enter a theological deliberation framed by dogma. The discussion would be in terms of absolutes set down by the Church and followed by faith; furthermore, religious decrees rarely consider the challenges of nonaffiliates. Members of the Church will decide the parameters of a religion, and it is neither the place of science nor nonbelievers to argue with representatives of God. When an institution is founded on the precept that mortals cannot always know the will of the arbiter of right and wrong, then there is little recourse for a "mortal" who wishes to confront the beliefs that develop from that sect. As Austrian philosopher Ludwig Wittgenstein had said,

> It can and frequently does, happen today that a person gives up a practice after recognizing an error on which it is based. But this happens only where making the person aware of his error is enough to dissuade him from his behavior. It does not happen where the religious customs of a people are concerned, and *therefore* there is *no* question of error in that case.[20]

Moral decisions are presupposed before the argument is made, and defying the belief can only result in an irresolvable shouting match between believers and nonbelievers. The question must be pushed back a step, becoming an existential exploration of whether God exists. If the answer is yes, then one can argue His will. However, the first point will remain unresolved because a believer and a nonbeliever will not agree, regardless of any presented evidence. Without agreeing on the existence of God, there can be no agreement on the morality He proclaims. As is probably evidenced, this discussion can

---

[20]Wittgenstein, Ludwig. Remarks on Frazer's The Golden Bough. In *Wittgenstein: Sources and Perspectives,* C. G. Luckland (Ed.), Ithaca, NY: Cornell University Press, 1996, p. 62.

only spiral further away from stem cells before it will ever come back. Suffice it to say that religions that oppose abortion do so for reasons routed in divinity and not, although not mutually exclusive, rationality. As religious scholar and ethicist Paul Ramsey puts it, "the value of human life is ultimately grounded in the value God is placing on it."[21]

Biology in this case, unlike religion, is open to interpretation. Scientists may have difficulty admitting that its definitions are uncertain, but disagreement over the genesis of life has been publicized by the abortion debate. At the heart of this debate is the moral status of embryos and whether they are equivalent to human life. Put another way, does a potential person have the same moral value as a person, and is it necessary to draw a line between the two? Biologically, however, whichever side can best appropriate scientific knowledge to support their claims is the group that will "win." Already, the deficiencies in this approach are emerging, in that evidence should induce the interpretation, not vice versa, but the evidence used by both sides is used compellingly.

The pro-life stance is rooted in the notion that "respect for persons" ought never to be compromised without adequate justification, which is a founding principle of bioethics. Terminating a life without the fetus' ability to defend itself is a gross violation of that principle. But, what convinces the antiabortion groups that the life is human? Beginning at conception, this group might answer, the human being has a unique genetic imprint. Furthermore, the embryo's genome will remain unchanged from the date of conception to the last day of its life. Granted, spontaneous mutations may occur, but the point is that an embryo's DNA is as unique and dynamic as that of any other human on the planet; the combination is just a little bit younger.

Although defining life in terms of a unique genome comes dangerously close to genetic reductionism, genetics is not the only determinant. If the pro-life movement were to rely on this definition, then shedding skin cells, hair, or any other cell in the body would be tantamount to murder. It is not enough to have unique genes (twins excepted), it is what happens when the

[21]Ramsey, Paul. The Morality of Abortion. In *Life or Death: Ethics and Opinions,* Daniel H. Labby (Ed.), Seattle: University of Washington Press, 1968, p. 71.

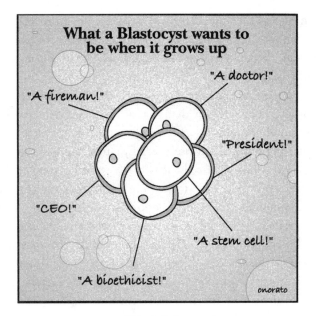

Figure 5-4.

23 chromosomes of the egg join with 23 chromosomes from the sperm—there is a potentiality within this newly formed cell to become fully human, which affords it the moral worth equivalent to a human. There is no natural potential within a skin cell to become an entire organism despite its genetic blueprint; indeed, after a few cell divisions, there is no natural potential for a single cell from a fertilized egg cell to become a human being (from totipotency to pluripotency to full differentiation). Opponents to this argument will criticize "potentiality" because they wonder why no preference is given to the egg or sperm, which also have substantial potential; the defense lies in the inviolability of conception. At conception, the events necessary to functionally confer personhood occur; the mechanisms that fuse the cells and bind the chromosomes, as well as initiate the cell division, commence. A sperm will never become a full-grown human being without that initiating event, and its fate is dependent on the natural circumstances that may or may not put it in contact with an egg. Furthermore, even if placed in contact with an egg, there is no guarantee that it will penetrate the ova and fer-

tilize it. An egg, on the other hand, can be stimulated to grow toward an embryo state without fertilization; however, it never has the potential to become a human because this anomalous process, parthenogenesis, ceases well before a heartbeat can develop.

As with the pro-life movement, the pro-choice supporters also rely on biology to defend their claim that women have the right to choose abortion or not. Autonomy, another foundational principle of bioethics, is the woman's guiding rationale. More precisely, since the embryo is not a human being, but a cluster of cells, it is merely another part of the woman's body. To validate this argument, the onus lies on proof of the nonhumanness of the embryo. Furthermore, critics contend that if the embryo is morally equivalent to a human, then the woman's decision over its fate is irrelevant. Bioethicist John C. Fletcher summarizes the view in his report on stem cells to the National Bioethics Advisory Commission: "The objection to this choice is that the woman abrogates this exercise of autonomy by the evil of the act of abortion and becomes, in effect, a morally defective decision maker."[22] What then makes something human in the eyes of the pro-choice movement?

Simply put, an embryo does not have those characteristics that make it distinctly human, so, the pro-choice argument goes, and until it does, it should not enjoy the moral status of a member of the human race. Regardless of its genetics, it does not have the basic biological qualities to merit precedence over a woman's right to choose. Until the embryo has specific physical and mental capacities, it is distinct from humans. In terms of the physical, these traits can be broken down into two categories, organs and form. Without all of the organs that humans develop before birth, the embryo does not function like a human. Furthermore, the appearance of an embryo as a mass of cells cannot be equated to a human; thus, without the form of a human, it cannot be considered one. There is an argument to be made here that the organs and form are assumed during the third trimester of development, so these criteria become moot. Two points should be directed at this critique from the pro-choice stance; the first is that many pro-

---

[22]Fletcher, John C. Deliberating Incrementally on Human Pluripotential Stem Cell Research. In *Ethical Issues in Human Stem Cell Research*. Volume II. Commissioned Papers for the National Bioethics Advisory Commission, 2000, p. E-15.

abortion activists will agree that this level of development is far enough along that an abortion would be unethical. However, those who support third trimester abortions often do so on the grounds of viability; if the fetus is not viable without the aid of medical technology, then it has not developed adequately enough to be human, particularly if it lacks the mental capabilities associated with human beings.

This second category, mental capacity, is very important to this debate. According to the pro-choice stance, sentience is critical to enjoying the moral status of a human. Although many believe that the capacity to reason is distinctly human, a baby is too young to be measured for rationality, although it is assumed that the child will develop some rational capacity. Opponents to this criterion often cite that animals have the awareness that confirms sentience, but it must be remembered that this argument does not stand on its own; rather, it is placed alongside those physical criteria previously mentioned. Again, however, many people who support the right to abortion will back off if these biological criteria are met prior to birth. If the fetus is viable, both physically and mentally, then it is human enough to have the corresponding moral status (as evidenced in the value they confer on premature babies and children born of caesarean section).

The leniency of the pro-choice supporters falls on the latter end of development. Because stem cells deal with the earliest two weeks of development, when they exhibit none of the biological characteristics that might be considered human enough by lenient abortion rights activists, the pro-life and pro-choice factions are strongly at odds over the research. The issue seems irresolvable given that both sides have as much evidence as they need (or want) to support their views. Very adroitly, scholar Ronald Dworkin explains the difference: "One side thinks that a human fetus is already a moral subject, an unborn child, from the moment of conception. The other thinks that a just-conceived fetus is merely a collection of cells under the command not of a brain but of a genetic code, no more a child, yet, than a just-fertilized egg is a chicken."[23] However, Dworkin continues his critique and observes that most liberals do not really believe that an embryo is just a collection of tissues, while few conservatives believe that an embryo is entirely

---

[23]Dworkin, Ronald. *Life's Dominion: An Argument about Abortion, Euthanasia, and Individual Freedom*. New York: Vintage, 1994, p. 10.

human: "very few people—even those who belong to the most vehemently anti-abortion group—actually believe that, whatever they say."[24]

Although extremists will not bow to compromise, treading the middle ground, unlikely though it may seem, is not impossible. In an attempt to acknowledge both viewpoints, Ronald Dworkin frames the argument in terms of policy to advise lawmakers on how to legislate abortion. In his terminology, one of two measures must be adopted as a guideline to regulation, either "derived" or "detached" approaches. In the derived view of policymaking, the decision is extrapolated from the absolute protection of the rights of a fetus, and the detached view focuses on the degree of violation done to the *concept* of human life.[25] Either the government has to prohibit it, or regulate it. Prohibition is kowtowing to just one faction, whereas regulation, or the derived view is a concession, although it may seem to be favoring the liberals. Returning to Fletcher's analysis, "Those who hold this view will believe the government ought either to prohibit abortion for this reason or, as the government has done in Roe v. Wade and elsewhere in the states, to regulate it by law. The most liberal view of abortion would insist on minimal regulation of abortion and maximal protection against intrusions on a woman's choices. Those who take the middle way will permit abortion but regulate it carefully by law. Prohibitions of abortion are appropriate when the fetus is viable, except in situations where abortion will avert threats to the woman's life or health."[26] This view is ambitious, as it attempts to do more than just find a common ground for debate. Choosing between derived and detached views also tries to erect the foundation for abortion policymaking in the United States, a democratic nation hosting an immense plurality of views. Yet, it also attempts to move the issue away from the moral status of the embryo, which will always be a limitation to this methodology in terms of bioethics, since moral status will remain at the heart of the matter.

The previous set of suggestions is useful for policy debates, but in terms of dealing with abortion outside of a legislative framework, the question still

---

[24]Ibid., 13.

[25]Ibid.

[22]Fletcher, John C. Deliberating Incrementally on Human Pluripotential Stem Cell Research. In *Ethical Issues in Human Stem Cell Research*. Volume II. Commissioned Papers for the National Bioethics Advisory Commission, 2000, p. E-28.

requires deliberation. Whether the embryo has moral equivalency to a human being is an all or nothing proposition to each side when it comes time to argue. However, John Robertson believes that this type of thinking must be abandoned in terms of scientific research because of the wavering that individuals on both sides exhibit when pressed on particulars: "Special respect but no rights for embryos makes sense if one views the underlying ethical and policy question as one of demonstrating respect for human life. If the embryo is too rudimentary in development to have interests, it may nevertheless be a potent symbol of human life."[27] Acknowledging that these cells are not completely human, yet represent a degree of humanity, it may be best to consider them "a potent symbol of human life."

Although the explanation above discusses the general aspects of the abortion debate as it relates to stem cells, the specific issues center on the different means of deriving these cells from embryos. There are four sources of stem cells that frame this debate. Although they may seem similar in scope, each deserves its own ethical analysis. These sources are: aborted fetuses (embryonic germ cells), in vitro fertilized embryos that are not going to be implanted, in vitro fertilized embryos created for research, and embryos created by somatic cell nuclear transfer–cloning.

In the first case, aborted fetuses, the debate has primarily focused on the relation of research and how it might influence the decision to abort a fetus. Antiabortion sentiment strictly forbids the act, but is the use of cells derived from aborted fetuses unacceptable if the decision to abort is unlinked to the decision to use cells for research? This question is centered on the notion of causation and complicity. In the eyes of those opposed to the act, regardless of the reason for the abortion, there are two levels at which using the fetus' cells for research violate their moral code. At the first level, the causative argument, the potential for research applications may have caused the woman to choose abortion. At the second, by using the cells for science, the investigators are supporting the act of abortion, thus their complicity with the practice is seen as furthering the violation. However, if there is no inducement involved, and the decision to abort is made prior to designating that the tissue may be used for research, then, the pro-choice groups would argue, these circumstances alter the relationship. Bioethicist John A. Robertson, who has thoroughly explored the relationship between causation and complicity, puts it this way:

> Under a causative theory of complicity, neither derivation nor later use of ES cells from abortions that would otherwise have occurred would make one morally complicit in the abortion itself because there is not reasonable basis for thinking that donation of tissue for research after the decision to abort has been made would have caused or brought about the abortion. Thus persons who think that induced abortion is immoral could support the use of fetal tissue or ES cells derived from abortions as long as the derivation or later research or therapy had no reasonable prospect of bringing about the abortion, just as they could support organ donation from homicide victims without approving the homicide that made the organs available. To do so however, such individuals would have to be convinced that research uses of fetal tissue from abortion otherwise occurring would not bring about future abortions or in some way make abortion appear to be a positive praiseworthy act.[28]

In effect, the embryo would be discarded if not used for research, thus some value may be derived from the procedure if the tissue is salvaged to better humankind. Protest against the procedure loses some foundation if it relates solely to research, which by this interpretation is unrelated to the act of abortion; derivation of the cells does not lead to the "death" of the fetus, rather, its death leads to research applications. Thus, the only real problem identified by Robertson is when individuals oppose fetal tissue for research on principle.

Aborted fetuses represent a practical problem to ES cell research in that there are too few to support extensive investigations, and identifying potential donors should occur after the decision to abort has been made. An alternative, indeed the most popular source of ES cells, is deriving the tissue from excess IVF embryos. Unused IVF embryos exist because couples that hope to bear children by in vitro fertilization usually donate enough sperm and eggs to create multiple embryos in case the first attempt proves unsuc-

---

[28]Robertson, J. A. Ethics and Policy in Embryonic Stem Cell Research. *Kennedy Institute of Ethics Journal, 9,* 2, 114, 1999.

cessful. After implanting a single embryo in the mother's womb, the remaining embryos (usually seven or eight) are cryogenically preserved until such time as the IVF clinic discards (destroys) them. Since many of these embryos are destined for death or eternal cryogenesis (unlikely due to the impracticality of eternal storage), researchers have found little reason not to use them for science. However, unlike aborted fetuses from which the cells are extracted following "death," deriving ES cells from an embryo *causes* its "death." What mitigates this consequence in the eyes of supporters is the notion that the tissue would never have progressed to individual viability and birth. "Death" is an inevitability; thus, if they are to be discarded, the embryos may be given purpose beyond cessation by aiding humanity through research. Again, however, Robertson's notion of the causative theory of complicity arises if the decision to donate is made prior to the sperm and egg donors' decision to destroy the material. Choice, in this instance, is not simply in the hands of the woman, since it is not her body that hosts the cells. At this stage, the embryo is in a test tube and, biologically, the male and female donors are responsible for decision making.

Not all IVF embryos are equal. In the previous case, embryos were created for the sake of reproduction; that is, they were created so that at least one could develop to birth and live a healthy life. Since this issue skirts the notion of aborting fetuses going through natural development in the womb, the temptation arises to use IVF to create embryos for the purpose of research. Although procedurally the details in this matter are identical to the previous case, the difference lies in intentionality: creating embryos for procreation is not intended at all in this case, and the mitigating factor of gaining one life in exchange for the loss of the remaining embryos is lost. Granted, this argument is only valid in those cases where some moral value is ascribed to embryos, regardless of their method of creation. Applying another belief system, for instance, that which maintains that there is no moral value to a fertilized egg at any stage prior to birth, would allow the creation of embryos for the sake of research without moral qualms. However, in one of the most influential essays on the topic, George Annas, Arthur Caplan, and Sherman Elias explain this issue at a much deeper level:

> To create embryos solely for research . . . seems morally wrong because it seems to cheapen the act of procreation and turns embryos into

commodities. Creating embryos specifically for research also puts women at risk as sources for ova for projects that provide them no benefit. The moral problem with making embryos for research is that as society we do not want to see embryos treated as products or mere objects, for fear that we will cheapen the value of parenting, risk commercializing procreation, and trivialize the act of procreation. It is society's moral attitude toward procreation and the interests of those whose gametes are involved in making the embryos that provide the moral force behind the restriction or prohibition of the manufacture of embryos for nonprocreative uses.[29]

Although this may seem like a pro-life stance, it is important to note that the statement is made to draw a distinction between IVF embryos created for procreation and those created for research. At the May 2000 Whitehead Policy Symposium on Genes and Society: Impact of New Technologies on Law, Medicine, and Policy, bioethicist Arthur Caplan, one of these authors, made it clear that he saw no moral dilemma with using IVF embryos created for procreation, but unused by the gamete donors.

This issue is far from resolved, particularly because of the distinction between research and application. Currently, few ES cells are required for research, limiting the number of embryos required for these efforts; however, when the technology progresses to the stage of development where therapies become common, it may be necessary to create embryos to meet demand. With this in mind, Robertson lays the foundation for justifying this approach, while directly answering Annas and colleagues' criticisms of creating embryos for science. "Given the controversial and sensitive nature of creating embryos for research," Robertson states, "it is likely . . . that research embryos will be created only for compelling reasons, for example, when important research cannot be validly conducted with spare embryos. . . . Indeed, leading proponents of creating embryos for research agree that restrictions on the sale of embryos are desirable."[30]

[29]Annas, G., Caplan, A., and Elias, S. The Politics of Human Embryo Research—Avoiding Ethical Gridlock. *New England Journal of Medicine, 334,* 1331, 1996.
[30]Robertson, J. A. Ethics and Policy in Embryonic Stem Cell Research. *Kennedy Institute of Ethics Journal, 9,* 2, 124, 1999.

Perhaps a better solution, although the verdict is far from in on this option, lies in the last source of stem cells in this discussion. In this scenario, embryos are created by somatic cell nuclear transfer (SCNT), or more commonly, cloning. Cloning is very complex, and its use in this debate adds greater fuel to this fire because of the controversy over the two different modes of creating cloned tissue for the research. In 1997, cloning became one of the largest issues in biotechnology when a group of Edinburgh-based scientists successfully cloned what may have become the most famous sheep in history; everyone seems to have heard of Dolly, and everyone seems to recognize the excitement and trepidation that accompanies her creation. Using the same technique, humans can be cloned, which has caused a stir that resonates far beyond bioethical and scientific circles and into the fabric of society. Because many of the criticisms brought against cloning deal with introducing cloned humans into the world, they are less relevant in this discussion because the cloned organisms will never make it to viability. Instead, they will be used to create tissue for therapy and experimentation. Eventually, this methodology will become critical to the process of stem cell therapy, but not as a means of creating the precursors of life; rather, cloning technology will allow transferring the genetic code from a sick individual into already isolated stem cells so that the ES cells will not be rejected by the new host body. However, until then, cloning for the purpose of creating embryos remains an option. When broken down into its technical and symbolic particulars, cloning becomes almost identical to creating embryos via IVF for stem cell harvesting, except that there is no fusion of sperm and egg; upsetting the sanctity conferred to conception, it is no longer an initiating factor in the antiabortion characterization of life. According to many of the pro-lifers, conception marks the beginning of human life; thus, the concern, some think, is obviated when conception is removed from the equation. Despite this technical loophole, it is naïve to think that one biological mode of creation will be privileged in this debate, which centers on the moral status of embryos and the perception of life. Most likely, the criteria used to define life will expand in the eyes of opponents to ES research, rather than remaining static; indeed, the redefinition of life and similar shifts in biological understanding indicate how deeply biotech can affect society. Such concerns should explain why so many are uneasy over biotechnology, regardless of its promise. Despite the differences in technique, the same issues associated with IVF quickly arise, as in the previous reference to the *New England Journal of Medicine* article by Annas, Caplan, and Elias. Thus, to use this method as a source of embryos reflects some of the concerns described above, while magnifying others, particularly the role of parenting in society. Huxley's *Brave New World* expands its cautionary breadth and joins more modern, popular admonitions like the film *The Matrix,* in which the instrumentation of human life through "creation" extends beyond social programming to supplying raw biological materials. The complexity of this topic has stymied many experts, causing them to shy away from using this methodology until more scientific and philosophical research is performed to understand the implications of cloning. In its report on stem cell research, the National Bioethics Advisory Commission weighs the benefits of the research with the social concerns, and is left wanting by their complexity:

> One major distinction between IVF and SCNT embryos is that while the creation of embryos by IVF would only generate more embryos, generation of embryos by SCNT would generate a specific kind of cell that might be useful in treating disease by allowing autologous transplant of a specific tissue type. Thus, in balancing the moral concern over the creation of an embryo and the value to society of the SCNT embryo, the potential therapeutic uses of the resulting ES cells from SCNT embryos must be evaluated carefully. At the present time, insufficient scientific evidence exists to evaluate this potential; however within the next several years, such information should become more abundant.[31]

Although there are analogies to be drawn with cloning human embryos, the second mode of SCNT embryo creation has fewer parallels. To move away from the debate surrounding human embryos, scientists have considered using stem cells from other species, cows for instance, and implanting them with human DNA to "fool" the cells into closely mimicking human embryonic tissue. The suspicion is that the cell matter's pluripotency will allow it to differentiate into any species' cells, thus delivering the benefits of

---

[31]NBAC, 1999, 56–57.

ES cell research while obviating the abortion debate. Reports from private industry claim that such research is very promising, but it also conjures *Island of Doctor Moreau* imagery. In both cases of cloning, the potential misuses of the technology raises suspicion. Will stem cell research be the excuse for enhancement experimentation to create humans according to social norms and biases, and will the research joining distinct species (chimeras) lead to the creation of new, dangerous species, putting instinct and reason in conflict for the sake of human desires? Perhaps more than any other issue, cloning and stem cell research show how the ethical issues that surround biotechnology can spiral toward greater intricacy even with the noblest of intentions. That is not to say that the research should not proceed, but it should do so with great wariness and strict regulation, which in the United States requires self-regulation because federal funding restrictions have placed much of this experimentation within the confines of the private sector. Cloning is not supported by federal funds, and although stem cell research has been, it is only allowed when ES cell derivation does not use any government resources. Congress has forbidden the National Institutes of Health to finance research on embryos, thus the cells have to be harvested and provided by the private sector, although the government is expected to support research on a number of existing stem cell lines within the next two years.

V. Industry

Stem cell research is not the first type of human embryo experimentation that promised great scientific and commercial benefit while facing anti-abortion sentiment. Decades before human ES cells showed strong therapeutic value, a similar debate broke out regarding infertility. In many ways, the history of in vitro fertilization can be seen as an earlier chapter in the bioethical story that continues with stem cells and may next proceed to cloning. The similarities are reflected at three levels: embryo research, in vitro fertilization, and private industry. The discussion began in the late 1960s when scientists proposed that experimentation on embryos might lead to better fertility treatments. In 1965, R. G. Edwards, a physiologist at Cambridge University in England, and Patrick Steptoe, a private practice physician, had inspired this discussion by fertilizing an ovum in a petri dish as the first step towards their goal of bringing in vitro fertilization from theory to reality. As might be expected, there was also a strong response against any type of research that threatened the life of a fetus. In the eyes of supporters, however, this research might damage embryos in its quest for knowledge, but the ultimate joy associated with helping people overcome their inability to conceive easily outweighed any moral objections.

In 1978, Louise Joy Brown was born; she was the first child brought to full term via IVF. There was an immediate response for and against such research and applications. Formal philosophical opinions had been developing in Australia, Great Britain, and the United States for quite some time, which eventually led to the formation of expert panels to officially examine the concerns. Britain's Warnock report, so titled because distinguished moral philosopher Mary Warnock headed the commission reviewing embryo experimentation, provided one of the earliest and most powerful analyses. However, its conclusions did not support a strong position; rather, it chose the relativist notion that "in questions of morality, though there may be better or worse judgments, there is no such thing as a correct judgment."[32] The United States faced the issue in 1978 when Dr. Pierre Soupart of Vanderbilt University submitted a grant to the NIH for funding a project on in vitro fertilization. Seeing an opportunity to use this grant as a sounding board for the issue, the United States not only put the grant in front of the Department of Health, Education and Welfare's (HEW) Ethics Advisory Board, it also issued an entire report on the topic. After a number of political offices reviewed the report, entitled *HEW Support of Research Involving In Vitro Fertilization,* which favored such research, it met with political pitfalls and eventually faded away without an official review. Because it was never given the opportunity for adoption by authorized individuals, the former stipulations on embryo research remained. These rules, which went through various revisions and governmental departments between 1973 and 1975 (each revision seemed only to change the wording and none of the sentiment), forbade research on fetuses unless it would directly help or if it represented minimal risk to the fetus. Since "minimal risk" is very restrictive, this posi-

---

[32]Warnock, Mary. *A Question of Life.* Oxford: Blackwell, 1985, p. 96.

tion would act as a de facto ban against federally funded research on embryos.

The almost thirty-year-old political, scientific, and ethical debate over this topic caused an interesting phenomenon to arise in science. Because this research actually could be a service to individuals, entrepreneurial scientists decided to remove the practice from the public sector and into the private. The science was going to progress, and if the government would not fund it, then the people who wanted the procedure would pay instead. Clinics arose and persist today because of regulations that may have defeated their own purpose. If they were meant to ensure stronger oversight of the research, the short-sighted regulations instead pushed the technology into the one place where it was beyond the government's grasp: private industry. In many ways, the moratorium created the in vitro fertilization industry. In another sense, in the more recent era in which funds have been plentiful for biotechnology, the regulations have done the same for embryo stem cell research, although it may have been easier to move the research into this sphere given the economic environment surrounding biotechnology today versus the 1970s. While Congress, the NIH, DHEW, and other agencies impose and lift bans, the confusion merely presents more opportunities to the private sector, which proceeds largely unaffected by federal moratoria on research. However, in the case of stem cells, the companies performing the research are well aware of the need for regulations in this polemical area; thus, many have decided to formulate guidelines on their own.

Although a number of groups are investigating stem cells, two stand out because of their early anticipation of the science, their relationship, and their intellectual property portfolio. The first, Geron Corporation, founded in 1992, began operations to confront and master the biological process of aging. Growing a substantial intellectual property base, the company began to focus on telomerase, an enzyme that plays a major role in regulating telomeres, the repeated DNA sequences located at the very ends of chromosomes. Telomeres are thought to be "molecular clocks" that gauge the age of cells, thus, Geron reasoned, resetting the clock would stay the aging process. As their research progressed, scientists and officers at Geron realized that to understand telomerase, a telomerase-positive source had to be identified; that is, a cell abundant in telomerase needed to be isolated. This quest led the company to stem cells, and more importantly, to the University of Wisconsin's James Thomson and Johns Hopkins' John Gearhart. Both of these investigators were, and remain, on the cutting edge of stem cell research; however, they were limited in their ability to perform their research because they could not access federal funds. Geron wanted stem cell data and had the money, and Thomson and Gearhart were in search of funding—a mutually beneficial match. Geron provided funding in exchange for exclusive rights to the resulting intellectual property. It is interesting to note that the third tier of Geron's technology platform is somatic cell nuclear transfer (cloning), another highly contested technology. As discussed earlier, cloning is an important technology to ensure that a host body does not reject en ES cell transplant; thus, this intellectual property is critical to a company that wants to develop ES cell therapeutics. But, the question that should immediately present itself when one considers that Geron has a great stake in both cloning and stem cell research is: Why isn't there a tremendous uproar over this company and its research? It should then be followed with the question: What have they done right?

Geron is an interesting case, and an excellent model, but that is not to say that the company's future does not hold bioethical trials and tribulations. The industry watches with great interest to see whether the firm can successfully maintain its strategy, or if they will lack the dynamism to anticipate changing social attitudes. Regardless of the future, Geron's present-day strategy deserves great attention. In Chapter 2, Monsanto and its genetically modified foods offer the case of what not to do, but here Geron offers the case of what a company can do to act responsibly and avoid a public backlash. Paul Rabinow, an anthropologist studying biotechnology at UC Berkeley, has pointed out that comparing Monsanto and Geron is useful, but not entirely fair, as the latter does not currently have products that can meet public scrutiny. However, ES cell research is under a social magnifying glass, which was not initially the case for GM food, and GM foods really met their greatest opposition and publicity when GM variants were found to contaminate storehouses of grain that were said to be devoid of genetic modification. This scenario, coupled with a social distrust of European regulatory agencies, precipitated a backlash against the industry that brought concerns to a much more visible level. Activists now had a clear example pointing to

GM crops as unnecessary, risky products furtively finding their way into food supplies. Transforming the issue from one of just risk into one of risk and corporate power, opponents to these crops made GM technology a potential threat instead of an agricultural marvel. Conversely, stem cell research began at the height of offense based on a polemic that is not as foreign to public attention as agriculture. Whereas the GM industry began its course below the social radar screen, embryonic stem cell research has always figured prominently in society's crosshairs.

It is still important to note that there is a sharp distinction between the types of "products" that each technology will manifest; in the case of stem cells, there will be therapeutics delivered by healthcare practitioners to just those few individuals in need of extreme treatment, whereas GM plants will pervade a substantially larger population, healthy or sick, without healthcare intermediaries. In the case of the latter, the "product" is food, and altering it will affect the majority of the world.

The question that remains is what has Geron done right to continue its research and remain socially responsible? Interestingly, the company struck a balance between two seemingly conflicting approaches; they were simultaneously suggestive and responsive. The subtlety lies in which approach is chosen under each circumstance. Upon reflection, the only way to reach heightened social awareness is to nurture these two traits, and Geron works diligently to do exactly that. Listening to the critique that surrounded the technology, the company made a substantial effort to assess the ethical issues surrounding the research, and considered how to address them. However, they went one step further. Understanding that the critiques by academic bioethicists were salient, Geron also recognized that these notions were often too esoteric and impractical to reflect the concerns of the public, which is a critical yardstick in determining levels of socially responsibility. It became important to the company to heed both concerns, which is a difficult task, particularly given the plurality of voices that weigh in on the topic.

Although it may seem obvious, responsiveness, that is, rationally considering the criticisms and fears associated with a technology, is the best way to prove corporate social awareness. The alternative is the traditional method of either ignoring criticism or launching a PR campaign that either subverts opinions or drowns them out. Luckily for biotechnology and the world affected by it, this industry is far from traditional. Still, the temptation is there, and considering how many resources these companies have at their disposal, it might seem likely that a firm would launch a public relations campaign to garner public support. Furthermore, these corporations might use some of their money to lobby the government to validate the utility of such research. Why not use all of its power to sway those groups that might interfere with scientific and corporate progress? Geron's answer is quite simple: that would be damaging to the company and the public. It would harm the company because the opposition can tap into the same avenues that Geron might use, and it would be harmful to the public because choosing confrontation over dialogue would exclude the point behind biological research, especially if corporations want to be profitable: delivering therapeutics that aid populations. Thus, the company has taken an admirable stance, best exemplified by CEO Thomas Okarma's statement: "It is important for us to avoid influencing the public debate. Our technology has ethical implications, and for us to be an ethical company, it would be wrong for us or any other group to lobby for support based on convincing people that they ought to see things our way."[33] Instead, Geron makes itself available to those groups looking for information and clarification on the technology to ensure that it may still inform the debate without steering it. For instance, Dr. Okarma (Vice President of Research and Development at the time) put all of Geron's cards on the table when he testified at the request of the U.S. Government to answer questions regarding the company's research during hearings overseen by the United States Senate Appropriations Committee's Labor, Health and Human Services, Education, and Related Agencies Subcommittee. The subcommittee was considering funding stem cell research. At this hearing, Okarma explained the significance of the research, why Geron chose to support it, how the NIH might gain by supporting the research, what the ethical concerns are, the company's research goals, its intellectual property position, and the conclusions of the firm's Ethics Advisory Board.[34]

This last point, the Ethics Advisory Board (EAB), explains the proactive

---

[33]Interview with Tom Okarma.
[34]Okarma, Thomas. Testimony before the Subcommittee on Labor, Health and Human Services, and Education of the Committee on Appropriations, December 2, 1998.

position of the company; it also highlights the company's commitment to ethical research. The EAB was put together because Geron recognized that biotechnology and bioethics are indivisible even, if not especially, at the corporate level. As technology progresses, there will be as much need for EABs as there is for scientific advisory boards. Although it may seem that an EAB and an institutional review board (IRB) are the same thing, the former is concerned solely with the ethics of corporate operations, and the latter divides its attention among many, if not all, corporate goals. An EAB advises the entire company and all of its operations (e.g., R&D, marketing, etc.) from a higher vantage point, thus advising on the "big picture," not particular experiments. Of course, the advice applies to individual experimentation, but the goal is greater: to ensure that bioethics assumes a substantial role in company decisions. In this spirit, Geron has put together a team of bioethicists, whose role in the firm is to act as objective advisors regarding the company's research, and the following statement was presented during Dr. Okarma's testimony to the Subcommittee to explain this very point.

A Statement on Human Embryonic Stem Cells by the Geron Ethics Advisory Board

Background:

Human embryonic stem (hES) cells have been derived and maintained in tissue culture. The hES cells are isolated from donated preimplantation embryos (technically, form the inner cell mass within the blastocyst) produced through In Vitro Fertilization for clinical purposes. Institutional Review Boards (IRBs) have approved the research. All the cell research is conducted within the guidelines of the 1994 report of the NIH embryo research panel and the 1997 National Bioethics Advisory Committee report.

What makes hES cells unique and important is that they are:

1. Pluripotent
2. Self-renewing
3. Expressive of the enzyme telomerase
4. Normal in chromosomal structure

Because of these features, these hES cells have the potential to make distinctive contributions to:

1. Understanding developmental biology (e.g., how tissues differentiate)
2. Pharmaceutical research (e.g., drug discovery and testing)
3. Transplantation medicine (e.g., generation of heart muscle cells, bone marrow, etc.)

There is no intention to use the hES cell lines for cloning a human person, transfer to a uterus (they could not develop into a fetus), or generate human–human or human–animal chimeras (mixing cells of different individuals or species).

Statement of the Geron Ethics Advisory Board:

An Ethics Advisory Board, whose members represent a variety of philosophical and theological traditions with a breadth of experience in health care ethics, was created by Geron Corporation in July 1998. The Board functions as an *independent* [italics added] entity, consulting and giving advice to the Corporation on ethical aspects of its work. The process of the Board's deliberations has included sessions informing the Board of the scientific, technical, and product development work of the Corporation. Members of the Board have no financial interest in Geron Corporation.

The Geron Ethics Advisory Board is unanimous in its judgment that research on hES cells can be conducted ethically. In order for such research to be conducted ethically in the current context, some conditions must pertain. In addition, further public discourse will be needed on a range of ethically complex questions generated by this research.

1. The blastocyst must be treated with the respect appropriate to early human embryonic tissue. Members of the Board are unanimous in taking a developmental view of the moral status of the developing human being. However, we also hold that, as *human* tissue, the blastocyst is to be treated with moral seriousness. Rsearch use of the blasotocyst requires justification; that justification is found in research that aims ultimately to save or heal

human life. This means that such tissue is to be used only when there is an overriding good to be derived from the research. In the view of the Board, the three purposes of research (understanding developmental biology, pharmaceutical research, and transplantation medicine) qualify as such goods.

2. Women/couples donating blastocysts produced in the process of In Vitro Fertilization must give full and informed consent for the use of blastocysts in research and in the development of cell lines from that tissue. The consent process must be undertaken with care, in recognition that donors undergoing In Vitro Fertilization are often vulnerable. Donors should understand the potential market implications of the research, and should be advised as to whether or not there are any proprietary rights in the tissue.

3. The research will not involve any cloning for the purposes of human reproduction, any transfer to a uterus, or any creation of chimeras.

4. Acquisition and development of the feeder layer necessary for the growth of hES cell lines in vitro must not violate accepted norms for human or animal research.

5. All such research must be done in a context of concern for global justice. One of the reasons the Ethics Advisory Board supports this research is its potential to contribute to widespread accessible medical interventions to alleviate human suffering. Accordingly, in the development of this research and its applications, attention must be paid to how technologies can be developed and utilized fairly for all people.

6. All such research should be approved by an independent Ethics Advisory Board in addition to an Institutional Review Board.

This analysis applies only to the isolation of hES cells from in vitro fertilized blastocysts. The Ethics Advisory Board has not yet considered the implication of emerging research on other stem cells isolated form fetal tissues.

Karen Lebacqz, Ph.D., EAB Chair
Michael Mendiola, Ph.D.

Ted Peters, Ph.D.
Ernle Young, Ph.D.
Laurie Zoloth-Dorman, Ph.D.

There are limits to an EAB, as critics are quick to point out. For instance, they are advisors, not officers of the company, thus the company reserves the right to adopt or deny the EAB's recommendations. Although Geron's scientists and administrators highly regard the advice of the EAB, Karen Lebaczq, its chair, does recognize this limitation as she explains in an article in the *New York Times Magazine:*

> "They are perfectly at liberty to ignore all our advice." Further, as Lebacqz points out, she is not free to discuss certain aspects of the research. "Early last summer, they brought us a piece of research that they were going to fund. Several members of the board raised objections, so they decided not to pursue that particular line of research." What was the research under discussion? "I'm sorry, I really can't [say]," she said.[35]

Although some view this as a defeat for ethics due to the lack of disclosure, it is quite the opposite. Research is constantly kept secret because of corporate strategy, thus it was not the ethics that prevented disclosure, it was the sensitivity of the science. More importantly, when the EAB objected, the company did not pursue the research. This point is critical; ethics won over commercialism. Although the argument can be made that it may have been a realization that profits would not justify the expenditure on research, the important point is that the company recognized the significance of ethics to both operation goals and commercial viability.

Geron is not the only company working with this technology, nor is it the only stem cell company concerned about the ethics of ES cell research. Advanced Cell Technology (ACT), a company that shares history with Geron, is also a player in this field. Headed by CEO Michael West, ACT took the stem cell research debate across species lines, which raised another

---

[35]Hall, Stephen S. The Recycled Generation. *The New York Times Magazine,* January 30, 2000.

red flag in this ethical debate. A founder of Geron, West left the company in 1997 and moved on to Origen Therapeutics, a company dedicated to research on avian transgenics. While there, West maintained his passion for aging research that led him to found Geron, and when new research from ACT demonstrated the means of using cow cells in lieu of human ES cells, he took up the banner for stem cells once again. Dr. West is much more vocal and active in the ethical debate than Geron, which demonstrates his conviction regarding the technology. Almost a crusader for antiaging, Dr. West sees the enormous potential of stem cells as the moral imperative in this debate. As Stephen Hall of the *New York Times Magazine* explains:

> In many ways Michael West is the shadow impresario of the field. As founder of Geron Corporation, one of this decade's most closely watched biotechnology companies, and now as president and C.E.O. of Advanced Cell Technology, West has achieved remarkable success as a kind of merchant of immortality, selling the idea that stem cells and related technologies might someday completely revise the tables of average human life span. And he is so convinced that the promise of stem cells justifies a controversial strategy like cow–human nuclear transfer that he is happy to foster, if not force, a national discussion of this technology.[36]

In January 1999, Dr. West forced the issue in the most public forum to date when the National Bioethics Advisory Commission convened hearings on stem cell research. Showing up uninvited, the CEO of ACT presented testimony in which he stated that "The increased incidence of age-related degenerative diseases will likely lead to conflicts of economics, ethics, and aesthetics as we struggle to find a humane and practical means of treating the ailing."[37] West's continuing point is that individual suffering may be substantially alleviated by stem cell therapy, which, in his mind is enough reason to warrant dedicated, scientific investigation to achieve this goal. For instance, therapeutic cloning technology will be a necessary aspect of stem

---

[36]Hall, Stephen S. *The New York Times Magazine,* January 30, 2000.
[37]West, Michael D. Testimony before the Subcommittee on Labor, Health and Human Services and Education of the Committee on Appropriations, January, 1999.

cell transplantation to overcome problems of histocompatibility (the body's natural immune response that leads to organ rejection). The prevailing notion is that the potential good of this technology, the relief of human suffering, outweighs its potential harm.

To say that Michael West is the leader of this movement would be a disservice to him and the movement. ACT's scientists who pioneered the nuclear transplant technology across species have been very vocal in the debate as well. These scientists, Robert Lanza, Jose B. Cibelli, and James M. Robl, have joined West in formulating ACT's ethical strategy. Strategy may seem a poor choice of words because the group did not decide on a strategy per se; rather, they clarified their understanding of the science and their belief as to how it fits within society and began to evangelize for it. They carefully chose their allies, and they found a cause that extended beyond the interests of ACT to unite these powerful voices. When seventy members of Congress wrote a letter urging a ban on federally funding stem cell research, Lanza took it upon himself to write a letter to the prestigious journal *Science* urging the Department of Health and Human Services to provide federal funding to the research.[38] Citing the unparalleled benefits of this research, Lanza and ACT recruited a host of famous, accomplished scientists to cosign the letter. Indeed, of the 73 authors, 67 were Nobel laureates. Interestingly, Lanza joined ACT just four days before the letter's publication, showing the kindred passion for the technology that he shares with West.

ACT's approach can backfire if it is not adequately tempered. The company experienced this in November of 2001, when it announced in both the electronic journal *The Journal of Regenerative Medicine* and the popular science magazine *Scientific American* that scientists at the firm had cloned a human embryo. Emphasizing the point that the goal was therapeutic (primarily to generate stem cells), the announcement was met with great scrutiny. From the scientific to the political community, ACT's achievement generated much the opposite reaction than it had hoped. Ian Wilmut, who led the research team that cloned the sheep Dolly, claimed that the experiment was far from successful (the cloned cell divided no further than the capability of some unfertilized eggs), while U.S. President George W. Bush registered his

---

[38]Lanza, R. et al. Science over Politics. *Science, 283* (5409), March 19, 1999.

own response, explaining that the event, and all cloning, was "morally wrong." Adding their voices to the clamor, many religious, antibiotech, and antiabortion groups aired their objects in the public sphere. Others claimed that ACT's announcement was no more than a publicity stunt aimed at raising capital. Congress promptly responded to this discord by introducing a motion for a six-month moratorium on any research on cloning human beings. Although the legislative verdict remains uncertain at the time this is being written, it is certain that ACT's pronouncement has successfully inspired a heightened, nationwide dialogue on cloning and stem cells.

It would be incorrect to conclude that encouraging public discourse is a mistake. Instead, drawing the public eye is a two-edged sword—very effective when aimed correctly, but to be wielded carefully. A company should not begin a dialogue with the public under the assumption that the company's view is the correct one, even if they are morally right. Instead, a true, open dialogue should be established. At the same time, ACT was not entirely at fault; the technological tools to clone have existed for quite some time and a bill had entered Congress to ban cloning well before this announcement had been made. If these were the only circumstances in this dialogue, then they may have been mitigating; but, as is often the case, they were not. Releasing the information in the wake of the September 11th terrorist attacks on the United States, ACT met with harsh criticism for its timing. While the country reeled over the decimation of the World Trade Center and the lives it contained, anthrax-laced letters also consumed the public's attention. Critics picked up on this, accusing ACT of trying to slip the experiment past those who might object. Stained by the accusation, whether true or not, the company pushed the boundaries of this strategy of forcing public discourse, and the repercussions of their decision may extend beyond the company to the entire stem cell industry.

Despite the cloning debacle, ACT continues to invite the public to assess the technology. Convinced that this dialogue has to take place in as public a forum as possible, the company had used the Internet to help convey this message. In contrast to Geron, whose website is noticeably devoid of ethical discussions, Advanced Cell Technology's Press Release and Recent Publication web pages[39] list a number of articles and links that lead to the discussions of the company's views, as well as informational sites regarding the de-

bate. This is not meant to imply that Geron has not shared its ethical reflections with the public; instead, Geron encourages its ethical advisers to publish the insights they have gained from observing the company, and inform
both academic bioethicists and the public through independent publications. In an effort to consolidate these views and ease their accessibility,
Thomas Okarma and the EAB members contributed to a special edition of
the premier bioethics journal, *The Hastings Center Report*.[40] In this special issue dedicated to stem cell research, the advisors presented their own interpretations of the discussion, independent of Geron's views, to inform the
bioethics community of the types of information that Geron was addressing.
Efforts such as these represent their dedication to academic freedom, which
allows EAB members to do their part to convey their insights as advisory
notes to the biotechnologists performing human embryonic research. Although both companies are still developing the technology, they recognize
the importance of the ethical debate. Their different approaches, Geron's in-
troverted approach and ACT's extroverted approach, both have merit, but,
more importantly, they represent their respective corporate cultures by reflecting the core values of each firm. Neither group could have made this
critical step if they had not taken the time to seriously consider the importance of ethics to both science and their operations—a good lesson to be
learned by the whole of the industry, not just stem cell research companies.

VI. Recommendations

For companies engaging in stem cell research, or any embryonic tissue research, some of the initiatives taken by ACT and Geron represent good
starting points for addressing the ethical issues facing the field. Although
each company must decide on its own strategy, one highly advisable path
rests in the self-reflection that aided these two companies in determining
their approach. At first, this suggestion may seem too vague or even unnecessary, but it is perhaps the most necessary step in developing a firm's core

---

[39] *http://www.advancedcell.com.*
[40] *Hastings Center Report, 29* (2), Mar–Apr, 1999.

values, which all policies should reflect. In particular, when dealing with a technology so closely linked to abortion, the temptation to not even ask questions regarding the morality of a corporation's policy must be avoided. The task ought not be thought of as a mental exercise when there is a clear goal, which is to internally and externally relay the corporate stance on the issue. Critical to this task is understanding the types of associations that will inevitably be made with the organization. There is no avoiding the link that the business will have to abortion, thus presenting this connection either proactively or responsively must be well considered. Because the most oppositional groups will claim that any organization performing stem cell research supports abortion, the researchers need to understand where they stand on the issue and to what extent this argument of complicity is valid. For instance, if a corporation believes that there is no problem with abortion in any context, then, at least, its internal views and policies should reflect this stance. However, if the corporation does not believe that it ought to engage the debate outside of the realm of science, then that must be clarified internally and externally, as well. More precisely, a corporation may not want to be linked to the pro-choice movement because its values or attitudes are neutral or opposed to a woman's right to choose. It may, instead, take the position that the social debate over abortion and family planning are not close to its ideals, and the justification for the corporate goals is explained in the company's commitment to performing ES cell research only in the context of IVF embryonic tissue because these embryos are destined for destruction.

Perhaps the most important question that faces the industry is what represents an "acceptable" source of ES cells? That is, what source of embryonic stem cells offers the best material, while causing the least offense? Given the choices—"extra" IVF embryos created for reproduction, aborted fetuses, IVF embryos created for research, and human or chimeric SCNT cells—the consensus among ethicists places the first two at the top of the list. Although certain types of research may require the latter two, they ought to be avoided because of the tremendous ethical baggage that they carry. There is a great deal of unease associated with creating "human life" or its precursors when almost immediate death is part of the plan. Emphasis should be placed on immortalizing stem cell lines from IVF or aborted fetus tissue so that the prospect of creating embryos for research is obviated. Until the technologies are perfected and ready for clinical trials and medical application, a process that would obviously require using SCNT, creation and cloning should be avoided. When the technologies are perfected, the discussion should be continued, adding to the knowledge already accumulated, but in this case, the SCNT technology should focus on pluripotent cells, which are used for tissue transplantion, rather than totipotent cells, which can develop into human clones if brought to term.

The decision to use one of the two ES cell sources is not as simple as choosing one or the other and proceeding with research. Quite the contrary, the ethical considerations surrounding IVF embryos created for research and aborted fetuses still require a considerable amount of deliberation prior to initiating research. Since the National Bioethics Advisory Commission has deemed these two sources acceptable, their recommendations offer excellent guidelines that can be adopted by the scientific community. Granted, the goal of NBAC is to outline the parameters by which research can receive government support, these boundaries are put in place based on careful ethical consideration; the extent to which a company chooses to adopt them, given the freedom that the private sector enjoys due to its lack of federal restraints, is up to the company and should reflect its core values. In both cases, NBAC's concerns center on the familiar concept of informed consent (see Chapter 3 for an in-depth discussion) on behalf of both gamete donors in the case of IVF and of the mother in the case of an aborted fetus. In each case, the Commission recommends specific criteria that ought to be met prior to the use of tissue in research or therapy:

> Prospective donors of embryos remaining after infertility treatments should receive timely, relevant, and appropriate information to make informed and voluntary choices regarding disposition of the embryos. Prior to considering the potential research use of the embryos, a prospective donor should have been presented with the option of storing the embryos, donating them to another woman, or discarding them. If a prospective donor chooses to discard embryos remaining after infertility treatment, the option of donating to research may be

presented. (At any point, the prospective donor's questions—including inquiries about possible research use of any embryos remaining after infertility treatment—should be answered truthfully, with all information that is relevant to the question presented.)

During the presentation about potential research use of embryos that would otherwise be discarded, the person seeking the donation should

    a) disclose that the ES cell research is not intended to provide medical benefit to embryo donors

    b) make clear that consenting or refusing to donate embryos to research will not affect the quality of any future care provided the prospective donors

    c) describe the general area of the research to be carried out with the embryos and the specific research protocol, if known

    d) disclose the source of funding and expected commercial benefits of the research with the embryos, if known

    e) make clear that embryos used in research will not be transferred to any woman's uterus, and

    f) make clear that the research will involve the destruction of the embryos.[41]

In federally funded research involving embryos remaining after infertility treatments, researchers may not promise donors that ES cells derived from their embryos will be used to treat patient–subjects specified by the donor.[42]

The difficulties in adopting these guidelines lie in the operational dissociation between sources of stem cells and research on stem cells. That is to say, infertility clinics provide the cells, whereas biotech companies perform the research. Critical to operating ethically, the firm performing the research must audit the suppliers of embryonic tissue to ensure that the tissue and

---

[41]National Bioethics Advisory Commission. Recommendation 5. *Ethical Issues in Human Stem Cell Research.* Volume I. September, 1999, p. 72.
[42]Ibid. Recommendation 6, p. 73.

donors are treated with the respect set forth in these guidelines and other policies held by the company.

So far as they are relevant, these recommendations for IVF should also be adopted for cadaveric fetal tissue. Of paramount importance is the concept of affording as much choice as possible before presenting the option for research use of the tissue. The amount of pressure placed on a woman who is in the midst of contemplating abortion is incredible, and that state ought not be exploited in an effort to aid research. For instance, a woman may be undecided, but when a researcher mentions the tremendous benefit that can be gained by donating the fetus to research, this information may tip the scales at a time when a woman is too vulnerable to rationally decide. All alternatives should be presented, and the woman should be provided whatever counseling and information she requests. After all options are presented and considered, if the woman chooses to go forward with the abortion, then, and only then, may she be presented with the option to donate the tissue to research.

The final recommendation lies in the formation of an Ethics Advisory Board. Although many companies have Institutional Review Boards (IRBs) due to legal considerations, or because they are progressive in nature, they may prove adequate for standard corporate functions, but should not be considered equivalent to an EAB. An EAB should not be limited to overseeing one experimental series at a time; rather, it should influence every aspect of the company, as every aspect of the company reflects the nature of the research being performed. From R&D to marketing, the ethics of the research can taint each division if not adequately addressed, and what would have been a successful product launch may turn into a disastrous business school case. The implication is not that ethics will always trump corporate decisions, although this would be the ideal, socially responsible corporate model; however, bioethical considerations should be made as much a part of day-to-day operations as possible. The responsibilities that lead to restrictions in the roles of CEOs, COOs, and other officers of the company may inadvertently exclude this level of thinking; thus, an objective voice stemming from the EAB may be the best source for these ideas.

The capabilities of this group need to be clearly defined. If they are not to have veto power over R&D, for instance, there should still be some mode of lobbying that will ensure that their opinions are seriously considered. The important aspect of this criterion is defining what serious consideration is. In brief, the ideas raised by the EAB need to be discussed and debated with adopted counterarguments clearly representing corporate values. These foundational principles will help the company understand both its position in the debate and its corporate goals, each of which are critical to the firm's success. To reach this level of discussion, meetings with senior staff and the Board of Directors will guarantee the proper place of the EAB as more than just window dressing. The integrity of the group will also be ensured by allowing the advisers to publish critical articles on the technology, which will also aid in gauging public perception and bringing debate closer to resolution (so long as proprietary information is not compromised).

Some might argue that, ideally, the EAB should have no financial stake in the company, although this is not necessary. These individuals are providing a substantial service to the company, and if the EAB shares the corporation's vision that a bioethical company leads to a successful company, providing a stake in the company may be an incentive to the EAB members to better perform the responsibilities of their office. Indeed, the concern over "buying" bioethicists to act as apologists is unfounded and overly simplified. Bioethics is primarily an academic concern, and bioethicists against this practice are unnecessarily preventing themselves from both receiving a source of income and testing the efficacy of the ideas that they have spent decades developing. The point that this community tends to miss is that an "apologist" on the payroll does little good for the company; biotechnology firms are not concerned with appeasing the bioethics community by claiming that one of their number is on the side of business. Indeed, the bioethics community is far from any consensus on these issues, so it would be ludicrous to believe that finding a thinker who agrees with the company has any use. The utility of an EAB is in highlighting the issues that affect the public and offering strategies to lessen any harm that the company's research might be creating. An apologist would merely find excuses to avoid preventative measures and construct rationalizations, which would severely underestimate the critical and vocal capacity of public concern over biotechnology. Furthermore, an EAB that did not offer recommendations and identify specific areas of public concern would be completely worthless to any company that has the foresight to include such a group. In the worst-case scenario, a firm may hire bioethicists to offer recommendations, even though the corporation has little interest in ethics; the company would bring in these scholars to work out the issues so as not to bother the corporate officers with the details, but assure the administration that the issues were under control. Although this is far from the ideal situation in which companies are dedicated to ethics and hope to affect a change towards bioethical operations, it is better than hiding from the issues entirely.

As to whom should be on the board, that question focuses on the type of technology and the issues that surround it. In cases involving embryonic tissue, bioethicists who understand the history of embryonic research are critical, as are scholars familiar with the abortion debate. Legal scholars are also important since so much of this debate has taken the form of court decisions and discussions over the legal limits of federal funding. Because of their interest in the abortion issue, feminist scholars both in favor and opposed to abortion should be invited to join the conversation. The more scientifically aware the EAB members are, the simpler it will be to assess the issues, although this is not decisive; indeed, many of those with the greatest concerns know very little about the science, but understand that the technology is related to abortion, which is enough of an association to warrant protest. An important skill set for an EAB is the ability to integrate solutions into corporate operations. Either someone who understands regulatory affairs or quality control (or both) is well positioned to see how ideas and suggestions may fit into the company's operations. The greatest difficulty will not be identifying the issues—they are redundantly available in the literature—the greatest difficulty will be integrating the solutions into standard operating procedures.

Stem cell research, although promising, may be the most controversial biotechnological innovation of the past few decades. Because it is so deeply entrenched within the abortion debate, many people have reacted very strongly to discoveries that surround it. Touching upon questions that center on the sanctity of human life, it is not surprising that the groups working in this field proceed with great caution and sensitivity to the public's views. Although it is a shame that such controversial topics as abortion are

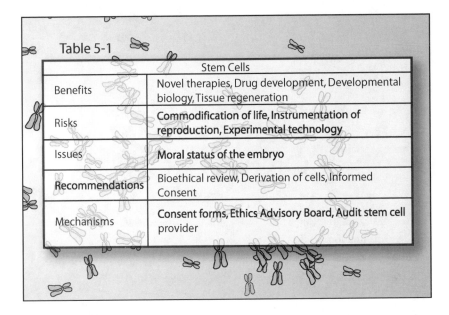

Table 5-1

	Stem Cells
Benefits	Novel therapies, Drug development, Developmental biology, Tissue regeneration
Risks	Commodification of life, Instrumentation of reproduction, Experimental technology
Issues	Moral status of the embryo
Recommendations	Bioethical review, Derivation of cells, Informed Consent
Mechanisms	Consent forms, Ethics Advisory Board, Audit stem cell provider

bringing the bioethical debate into the corporate context, it is promising that these discussions offer a foundation upon which constructive criticism and social responsibility will build a new legacy for biotechnology corporations.

6

CONCLUSION

Every great advance in science has issued from a new audacity of imagination.

—John Dewey, *The Quest for Certainty*

Readers familiar with bioethics have most likely discovered that this book is a departure from traditional works analyzing the tension between bioethics and biotechnology. Indeed, the usual elements of judging the morality of research have been replaced by offering recommendations that allow the reader and, more importantly, a corporation to choose responsible avenues of conduct. Rather than make moral judgments, these pages reexamine the moral relationship between biotechnology and society by including the oft-neglected factor of industry, which allows corporations, scientists, and other concerned people to view more accurately the need to balance bioethics with biotechnology. However, this text does not stray so far as to avoid framing and addressing an ethical dilemma. The problem in this case has shifted from the moral basis of research to the moral basis of technoscience. More precisely, biology by itself ought not be the singular focus of bioethics; instead, ethicists must examine how technology and science combine in the form of a corporate "product" before it penetrates the world, and they must do so with a deep respect for and an understanding of the business landscape.

By now it should be obvious that this is not a job for bioethicists alone. Ethics cannot be forced into corporate practice; the biotechnology industry

230

must first extend an invitation to this discipline. Until corporations have made the decision that bioethics is worth pursuing as both a humanistic and business goal, the transition cannot occur. The mental barrier is not the issue; corporations most likely believe that their core values propel them in such a direction, but without examining and critiquing their operations, companies will not be able to make this transition. In many ways, corporations are in the ideal position to force this shift, and the time is right to do so—before the course is chosen *for* them rather than *by* them. All that biotechnology needs to do is follow the lead of those companies that believe as strongly in ethics as they do in their technology, and then the pieces will begin to fall into place. That is easier said than done. The challenge lies on the shoulders of industry, which needs to take the lead by guiding and educating bioethicists, while asking the latter for guidance and education. A synergy between groups must be struck, whereby corporations relay the intricacies of business operations, and bioethicists relate moral and ethical theory. The tools to start are in these pages, but the dedication and commitment must come from those who read them.

The industrial view of biotechnology has placed science, and more precisely, technology, in a bureaucratic framework: progress is linked to organizational structures, either through corporate research or academic alliances. Thus, the drive for efficiency at all levels of industrial operations (administrative structures, research and development, etc.) joins technical practices and bureaucratic systems, forming a technocracy. Modern biology's ability to alter humanity's understanding of itself and society coupled with the systematic efficiency governing technological advance represent converging domains. The details of this bureaucracy are new to many bioethicists, and it is at this convergence that the ethical dimension must be included.

Because biotechnology corporations control the research on and distribution of these "products," it is in many ways the responsibility of these companies to decide the extent to which they will serve the public interest. On the whole, I side with companies due to the public health service that biotechnology delivers; thus, my bias falls on the side of technology and its promise. However, companies still need to proceed with caution. When examining these issues, I do not see unethical technologies; rather, I see potential areas of abuse questionably justified by technology's possibilities. Nothing so profound as biotechnology should be viewed as so beneficial that it should go unchecked. Yet, there are too many biases, not dissimilar from my own faith in science, that force individuals to unwittingly downplay the *potential* damage that technology might produce in favor of the cures and treatments it *might* bring.

Having acknowledged my bias, it has been the goal of this book to highlight the ethical concerns that face technology, regardless of which side I may favor. When I disclosed my sympathy to industry, I did not do so to justify its practices; rather, I did it to explain my attitudes toward bioethics with respect to corporations. Recalling the introduction, the point was made that a consulting bioethicist would do no good to a corporation by helping it rationalize its business practices. Instead, a bioethicist's first obligation is to serve the interest of ethics, which requires an objective presentation of the views weighing in on a debate. Opinions vary substantially across society, and that diversity is also apparent in the discipline of bioethics. There are some general areas of consensus, but on the whole, ethicists, unsurprisingly, are still human, and they carry many of their predispositions with them to work. In fact, that is what we expect of them. Without bringing their experiences to their subject matter, either from scholarship or observations of science, or even from debates with their layman friends, bioethicists would have very little to offer. What they do offer is a thoughtful critique of technology and its social implications predicated upon a much deeper reflection of the topic than most people have the time or energy to cultivate.

When dealing with industry, bioethicists have an opportunity to do more than publish and obtain tenure. The private sector contains unique sets of problems, each of which have to be addressed in a context often foreign to ethicists. The corporate world differs from the academic, and issues are transformed when ideas are placed in this context, entering the more practical realm rather than the theoretical. However, it is here where the concerns around biotechnology are the most important because technologies are rarely introduced into society through academic centers; furthermore, these issues are rarely answered exclusively through theory. Technology reaches society through corporations, and technology is at its most pervasive when it is distributed as a product. This commodification adds an entirely new dimension to the bioethical calculus, where it does not necessarily exclude bioethists from the equation; rather, it invites corporations to reflect on technology as profoundly as the bioethicist do. More importantly, the interests of both parties must be addressed if biotechnology's implications are to be handled responsibly.

With the background gained from the preceding chapters, it is now easier to explain which institutions are at play on the biotechnological stage, and how their interests affect their decisions, conduct, and each other. Anyone familiar with these topics will immediately recognize two characters at the nexus of biotechnology, while the third's often overlooked part has been augmented by the argument of this text. They are science, society, and industry.

Each of these institutions has its own goals and interests. Despite their actual complexity, viewing these institutions according to their prevailing ideologies simplifies the analysis and clarifies their motivations. Philosophers have struggled with the concept of "science" for millennia. Its ideal has always been truth or, more precisely, knowledge. Science's ideology is one that focuses on the unmasking of the natural world and understanding the phenomena we perceive every day. When it comes to questions of organisms, biology is the division of science that attempts to answer these questions surrounding the natural order of the world. In the abstract, then, the ideological goal of science continues to be the pursuit of knowledge, which is the interest it serves.

What then is the character, ideology, or goal of society? This is a much more difficult question, and it is almost more damaging to try to answer it in a paragraph than not answering it at all, although an attempt must be made to fully explain the present argument. There are many facets to society, and an equal number of pursuits, which makes bioethics' goal of trying to serve society increasingly difficult. However, for the purpose of discussion, only one aspect will receive attention: justice. Society's interests range from equity to peace to order and many other concepts in between. Taking these ideals into account, justice seems to address each of them in some tangential way, if not directly. Justice also comprises one of the four traditional principles of bioethics, which are often described as beneficence, nonmalfeasance, autonomy, and justice. Although the derivation of these concepts in bioethics is nuanced and complex,[1] it is not hard to see how the three other notions may be considered as degrees of justice, instead of singular concepts.

Interestingly, examining the interests of industry does not cause the same type of hermeneutic problems. Most people look upon industry and say that its only motivation is profit. A business, biotech or not, is in business to succeed, and to succeed it must bring in revenue. Actors within this framework have an obligation to maximize profit and minimize costs, which solidifies their commitment to the industry's ideological goals.

In a moment, some of the difficulties of looking at these institutions as solely motivated by their abstracted ideals will be explored, but first their conflicts require discussion. Science and society have always had some degree of disagreement regarding the conditions of their relationship. Whether science's pursuit of knowledge ought to eclipse the interests of society (scientists would argue that their work will ultimately aid society) remains less an issue than whether scientists know better than society. Authorities in biology, for instance, often take the stance that their knowledge affords them the better vantage point in social discussions, thus the layperson opposed to progress ought not weigh in on the debate. Put another way, science knows what is good for society like a parent knows what is good for the child. However, members of society do not want to be dictated to, and they may levy the criticism that the pursuit of knowledge is not the sole motivation of scientists. Careers, tenure, publishing, and funding are also very important issues to the scientist, which raises a degree of skepticism on behalf of many members of society. How does one know that a scientist has not confused his or her personal goals with his or her scientific project? Sequencing the human genome is a tremendous accomplishment, but many individuals are wary of the project given the ethical and legal issues that surround it. It is easy for a member of the Human Genome Project to claim that it serves the public good, but will the public remain foremost in an investigator's mind when he or she receives enough funding to maintain his or her research and to increase his or her salary? Through the "pursuit of knowledge," careers are made, board appointments are distributed, stock options are offered, and

---

[1]Beauchamp, T. L. and Childress, J. F. *Principles of Biomedical Ethics*. New York: Oxford University Press, 2001.

many other perquisites are awarded. To say that none of these benefits affected the investigator's decisions and motivations casts serious suspicions on the honesty behind the claim.

Despite this criticism, there is something to be said about the ignorance of the public. Most do not have PhDs, and many have only heard about biotechnology from the news, which is far from informative on the topic. However, ignorance has never halted critique, and more often than not ignorance fuels it. Some people have taken the time to consider the issues, and others have not. Some people have been guided by objectivity, while others have been led by ideology, although many do fall in between. Even more frustrating, some people have been motivated by their allegiance to society, while others have been motivated by their rampant self-interests. All of these phenomena make it difficult to say who ought to be making decisions about these issues; however, in a society distinguished by pluralism, everyone deserves their say.

Tensions are amplified when the profit motive enters into play, which can pit industry against both science and society. In the case of science, the debate over patents best exemplifies the issue. Corporations have to patent genes and other novel discoveries, but many claim that this practice conflicts with the interests of science because so much research is kept secret until a patent is issued. Science proceeds without the benefit of known information when a company keeps ideas confidential until it is protected. Interpreting patents in this way, members of society may doubt the possibility of altruism in the private sector simply because Wall Street's expectations guide so much of corporate thought. On the other hand, many advances can only be made by industry, and then, only if companies obtain the necessary resources to bring therapies to the public. It is costly to develop new drugs, and beyond product development, the overhead necessary to take a treatment from scientific validation to public distribution is considerable. No other institution but industry can do this, and as such, no other institution can deliver public health as effectively.

These institutions are not always at odds, but what becomes clearer in examining them according to the above description is that their ideologies can guide them in divergent directions. Still, there is a great deal of overlap, and that overlap needs to be explored and strengthened. More importantly, each must be considered as more than just singularly goal-oriented organizations. Although it would be ideal to have all of their ends overlap, it is also critical to examine each institution as a sum of its members. Stepping back a moment, it becomes obvious that each institution shares members from the other. For instance, members of society and the scientific community consistently participate in industry. Even more importantly, people in industry are part of society and the scientific community. Although it seems an obvious point, it does not seem too far-fetched to believe that when representing one of these roles, particularly that of industry, an individual may disregard how his goals as an agent of industry may conflict with his goals as a member of society. It is this scenario that has to be avoided.

Biotechnology has too great an impact on the world to become merely a tool for industry to meet quarterly expectations, and for the freedom that corporations need to operate, biotechnology firms must understand the importance of bioethical critiques, which are intended to help situate biotechnical science within society. It is too easy to fall into the role of corporate executive and forget about one's other obligations, given the pressures of operating a biotechnology firm; however, there is a very important, self-interest reason for corporate officers to address bioethics. By helping biology proceed responsibly—that is, when bioethicists encourage responsible and productive research—corporate representatives would do well to heed this advice because in serving the interests of society, bioethicists also serve every biotechnology practitioner. After all, working for a biotechnology company does not excuse someone from being a part of society. Although it is nobler to act ethically because one believes it is the right thing to do, for those who go unconvinced, perhaps this argument of self-interest will aid the acceptance of bioethics in the corporate world.

There are other reasons to bring society, science, and industry together through corporate efforts. Bioethics, biotechnology, and corporations, the groups that represent these institutions, each have a great interest in seeing therapies aid humanity. However, there is another institution interacting with these groups, that of government. Until now, there has been little mention of government, but it has a prominent place in bioethical discussions. The government's involvement with these institutions generally breaks down as follows: With respect to a democratic society, government represents it; with respect to industry, government often regulates it; and with respect to science, government often funds it. However, when private dollars are involved, government is often removed from the equation. By funding science, governments can do so on specific conditions, thus restricting research that may not be in the best interest of the sector it serves, society. When funded by venture capital or another private source, biotech companies are only limited by their respect for the law. However, in representing society, the government cannot turn a blind eye when issues are at the forefront of public consciousness, and for this reason, it behooves industry to pay attention to this delicately balanced relationship before legislators take steps to control a greater portion of scientific investigation.

A topical example, and one that will not go away in the near future, is cloning. Cloning human tissue creates a substantial number of concerns, but there are tremendous benefits should the technology proceed to its therapeutic objectives. However, almost every individual has an opinion on whether human cloning ought to be performed. Questions ranging from the value of a clone's life to the threat of eugenics hang over this technology. In fact, almost every bioethical concern can find a home in the cloning debate. In the United States, this issue has recently taken the political center stage, which promises to host it for quite a few years.

Cloning offers an excellent example of why bioethics are critical to industry because it highlights the very important point of power. Those who have power in this arena are expected to wield it responsibly, and if a more powerful group sees that a weaker group may mishandle the information, the more powerful group is charged with taking steps to control the technology. In this case, the concerns of the public are so profound that politicians had to respond. While Congressmen are still influenced by their usual interests and political leanings, they did what they thought needed to be done. In mid-2001, a bill was drafted and brought to the House of Representatives to ban cloning. It should be noted that this scenario is much different from the stem cell debates that have recently occupied Washington. With stem cells, the issue was whether or not federal funds could be allocated to stem cell research. With cloning, the issue is to make pursuing the technology illegal. This is the greatest concern that industry should have. If governments ban types of research, rather than regulate them, then companies are left with little or no recourse. They have no choice but to obey the law and relinquish any research agenda that implicates a banned technology. Industry should always push towards self-regulation, but it does not deserve that level of freedom unless it does so with an incredibly dedicated eye towards social responsibility. More precisely, it cannot do so without extreme deference to bioethics; it is the choice between self-regulation and no research at all.

Looking at how these interests have clashed over cloning explains how a worst-case scenario for industry plays out. The Biotechnology Industry Organization (BIO) has assumed the role of public speaker on the behalf of industry. As a governor of the industry, which includes a bioethics division, BIO is also the sector's lobbyist, and it had an obligation to fulfill those duties when cloning came to the floor of the House of Representatives. Favoring a less restrictive act, which would allow cloning research, instead of the alarmist bill that bans it altogether, BIO President Carl Feldbaum explained, "once the Congress starts criminalizing avenues of research, then we have a big problem."[2] Despite his clear religious influences, Richard Doerflinger of the U.S. Conference of Catholic Bishops concisely states the general social concern over corporate efforts at cloning and how the government must reflect them: "BIO has certainly given the impression that it sees no moral limits to this technology. As a result, there is a growing perception in Congress that any moral limits on the manipulation of life by these companies will have to come from outside."[3] In this case, science, society, and industry are at odds over this technology, and each can easily be interpreted as concerned with its own interests at the expense of the other. This is not to say that none of these institutions are concerned with the other; quite the contrary, but each group looks suspiciously upon the other. For instance, examining Feldbaum's statement above, the question as to whom "we" refers is critical. Given his record of including ethics within BIO, those familiar with the group would read the "we" as referring to all those who may benefit from technology, thus the whole of society. However, when there is passionate discord over the appropriateness of a research agenda, critics from opposing camps may look at that statement and interpret the "we" as corporations. If that reasoning continues, then furthering corporate interests,

---

[2]Usdin, Steve. Losing Goodwill. *BioCentury*, August 6, 2001, A7.
[3]Ibid.

interpreted as profit, corrupts any nobility that might emerge from the technology. Why would anyone want a corporation to control a contested technology if the risks involved are justified through potential revenues? With this scenario, one could only expect government to relinquish any claim to laissez-faire politics, and regulate, if not ban the technology. It would be the best way to serve society in light of the corporate and scientific pressures that appear to be at odds with it (see Figure 6-1).

One might counter by claiming that bills are not always made into laws, which is the likely case for the cloning bill when it reaches the Senate. Although President George W. Bush has stated that he would uphold a ban on cloning, it is likely that the Senate will reject it for many of the same reasons that convinced the California Supreme Court to act on behalf of "biotechnology" in the *Moore v. UC Regents* case discussed in Chapter 3. I do not mean to trivialize cloning, and although it is unlikely that it will be banned outright, the technology's legality is far from guaranteed. However, these legislative discussions affect the standing of a company linked to cloning, and if companies in this atmosphere assume an attitude as brazen as that of Advanced Cell Technology when it released its data on human cloning experiments (see Chapter 5), public outcry may force Congress to outlaw related research, or respond by introducing strict sanctions. It is not as deleterious as having a firm's platform technology become illegal, but being linked to a controversial technology that seems to pit industry and science against society does not aid the corporate agenda. As has been the case with some biotechnologies, particularly GM foods, corporate sponsorship from venture capital groups and other sources become scarce when a technology has been labeled as risky or unethical.[4]

Although legislation ensures to the highest degree that certain research will not take place without penalty, failed legislation and/or governmental debates still assign a stigma to a corresponding technology. Using stem cells as an example, a bill was never introduced to ban stem cell research, but heated debate continues even after President George W. Bush made his decision on whether to uphold or withdraw a ban on federally funding this research. Ultimately, he made a compromise by funding research on estab-

---

[4]Dry Season. *The Economist*, November 2, 2000. *http://www.economist.com/displayStory. cfm?Story_ID=413367.*

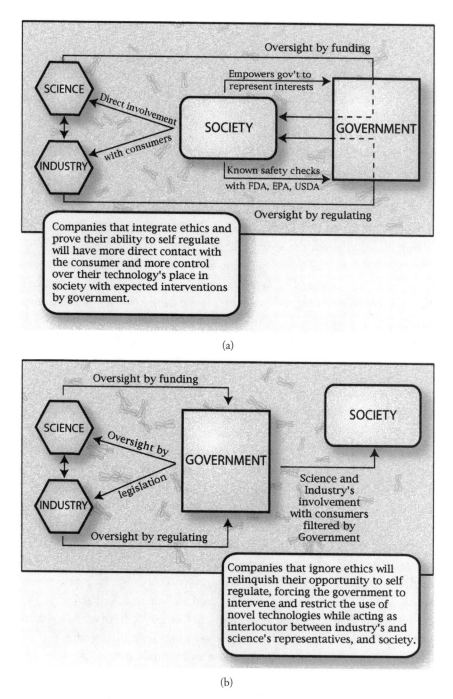

Figure 6-1 Society, industry, science, and government.

lished cell lines, but not for establishing cell lines; the latter would imply transforming embryos from potential human beings into experimental, biological matter. Absolving himself from complicity with destroying embryos, Bush also made it difficult for this subset of the biotechnology sector to continue innovating research. Private funding in this area is becoming difficult to obtain as investors try to distance themselves from an ethically questionable practice; thus, many stem cell companies were counting on federal funding opportunities. While some taxpayer dollars will support stem cell research, there will certainly not be enough to make up for investors who are made skittish by this technology's social baggage. Critical to their success, these companies must distinguish themselves in the ethical and policy arenas as much as they already have in the scientific.

Throughout this text, comments have been about the polemical nature of bioethics, within both the biotechnology and academic spheres. One may wonder how bioethics would be useful to a company performing stem cell research or contemplating cloning when the so-called experts cannot agree on acceptable standards of investigation. Inviting bioethicists or someone well versed in bioethical arguments to assess a corporation's operations and research is still very worthwhile. Such an "expert" will be able to delineate the objections towards research according to their severity. More precisely, reviewing the literature will reveal which viewpoints have gained general acceptance in the bioethics community and which have not. This gauge indicates to some extent, although not completely, how the public and policymakers would receive different research agendas. To be responsible, bioethicists must address public concerns, and, more often than not, bioethicists criticize the suspicious components of a technology in proportion to public sentiment. For instance, many people come out on the side of industry and science through their understanding of the importance of stem cell research. The incredible therapeutic potential offered by these cells is almost universally recognized; however, it is important to explore other areas of general consensus. Just because people realize the value of research does not mean that the same people accept the value as absolute. The following question should then be asked: are there conditions that the bioethics community generally accept as a best-case scenario for addressing ethical issues in stem cell research? If the answer is yes, then the first step in proceeding ethically would be to adopt or convincingly dispute these standards.

Ignoring these "consensus" standards can damage a company's reputation either by casting doubt on their altruism or by reflecting their interests as purely profit driven. Furthermore, the firm's commitment to ethics may be seriously questioned, compounding any public relations damage. A recent example captures this sentiment. At the Jones Institute for Reproductive Medicine, scientists recruited ethicists to "bless" their intentions to create embryos for research purposes. In Chapter 5, this book advises strongly against it, but scientists at the Jones Institute relied on other authorities in the field, claiming that stem cell creation was a moral imperative. However, shortly after this decision, *The New York Times* ran an article with the headline "Bioethicists Fall under Familiar Scrutiny."[5] A general consensus in the bioethics' discipline (there is never total consensus in any field) exists claiming that embryos should not be created for research purposes, and only those that are created for the purpose of reproduction, but are not going to be brought to term, should be used for experimentation. The Jones Institute, which often meets this requirement with its unimplanted IVF embryos, found ethicists who disagree. However, when asked the question who these ethicists were, the clinic's spokeswoman, Jane Gardner, answered, "When people signed on to participate in the ethics review, that was not with the understanding that their names would be made public. That wasn't part of the deal."[6] There are many reasons to speculate as to how this decision was made, and it is not difficult to guess from their writings which ethicists were asked to participate in this decision, but it is difficult to trust the intentions of the institute. Importantly, it was not a bioethicist, scientist, or corporate representative who brought this incident to public attention. Instead it was the popular press: *The New York Times.* When a journal with such wide distribution and authority focuses on a topic, the issue is no longer left to the specialists. It becomes a public, that is, societal, issue, moving bioethics from behind the curtain to center stage. However, one may ask how ethics helped

---

[5]Stolberg, Sheryl Gay. Bioethicists Fall Under Familiar Scrutiny. *The New York Times,* August 1, 2001.
[6]Ibid.

Figure 6-2.

the cause if bioethicists were consulted and *The New York Times* was left asking who they were. In this case, it did not help, but not due to a failing of ethics, but rather due to a failing of the Jones Institute.

The Jones Institute acted in a number of questionable ways that should be avoided, regardless of the profitability of their actions. First, by claiming, as many scientists who agree with creating embryos for research do, that it is a moral imperative to do this research, the group has revealed a very strong bias. Recalling the discussion over the interests of science just a few pages earlier, scientists are invested in their work, and often it is difficult to accept their "moral" reasoning, no matter how valid, as the basis of a decision when the same polemical decision results in their gain. In this case, the somewhat suspicious decision goes against the position of most thinkers, which invokes the second mistake. It is not unreasonable to assume that the Jones Institute ethics advisors were selected for their minority view that embryos should be created for research purposes. There is a scientific need to do this research, especially to study early embryo development, but the technology is still in its earliest stages and difficult to control. Scientifically, animal models and other manipulations are necessary before this type of stem cell experimentation gains greater moral acceptance. Thus, when science better understands the actual contributions, rather the therapeutic contributions, as opposed to the current conjectures, then the "moral imperative" to further this research will have a better foundation to support debate from both sides. By choosing "yes men" as their ethicists, scientists at the Jones Institute most likely presupposed a moral answer and then found the individuals who could give it. This strategy defeats the purpose of enlisting ethicists and introducing bioethics into corporate operations, which is why *The New York Times* very promptly ran a story doubting not only the intentions of the Jones Institute, but the utility of bioethicists within corporations. Despite this latter concern, there is a clear place for bioethicists in corporations, but it is contingent on their integrity. To do anything but be objective is a disservice to the field and to the corporation. And part of that objectivity is relaying all points of view, as well as their degrees of acceptance. It is ultimately the decision of the company to decide which path to follow, but when it decides to challenge the consensus, it is incorrect to do so and claim that it is "ethical" when it knows full well that most experts disagree. Furthermore, trying to displace culpability onto the shoulders of anonymous ethicists, rather than confronting the issue as a company, is a severe mistake. The ethicists are not responsible, but the company is. Ethicists inform a decision, and perhaps they do so very persuasively, but management will make the ultimate decision. That decision may reject or embrace the advice, but when the ethical basis for it is kept confidential, and when the "experts" are hidden, then who is to say that ethics had any bearing on the choice? When a scenario like this plays out, corporations do a great disservice to their stakeholders by arbitrarily holding bioethicists up as a moral shield. When a company or institute makes a business decision, it needs to stay true to that framework; a business decision is just that: business, not ethics. It will take a lot more to make them identical.

Interestingly, the Jones Institute halted its practice of creating embryos for research purposes almost as quickly as it made the decision to do so. When the scientist who began the research retired in January of 2002, the members of the Institute agreed to continue performing stem cell research, but only by accessing already-established cell lines. When asked whether political pressure or personnel changes inspired this shift in approach, Dr. William E. Gibbons, chairman of the department of obstetrics and gynecology, acknowledged the influence of both.[7]

It is at the level of institutional interaction that science and technology are strongly influenced by the structures governing their interface; thus, these organizations require attention in ethical debates. The question then arises as to how bioethicists can penetrate a debate that has assumed a new form, and what the form in question is. The answer to the second question informs the first. At the heart of scientific advance is the industrial vision of progress. More precisely, emphasis on efficiency has led to a technocratic framework in which science is practiced. Both in academia and corporations, bureaucratic systems are in place in the form of grant writing procedures, management, marketing pipelines, design controls, etc. The ideological dedication of these systems to technology development represents the end of these systematic means. It is here where ethicists have failed to take their necessary place in the life sciences because engaging bureaucracies is a daunting task. Sociologist Alvin Gouldner aptly describes the intricacy of these assemblages:

> The technical staff is governed by and subordinated to an officialdom which sets the goals, but which knows little about the technical process used to realize them. The technical staff is alienated from the ends; the officials from the means."[8]

Thus, individuals who know themselves to be moral assume that their conduct will be consistent with their character; however, corporate officers and scientists may be so caught up in their functional role that ethical conduct is not explicitly considered; it is neither part of the ends nor the means.

Despite the prevailing notion that bureaucracies limit functionality through restrictive "red tape," their prevalence among successful research centers speaks both to their efficiency in managing different divisions and to their permanent place among them. This technocratic model may seem im-

---

[7]Associated Press. Virginia Institute will Stop Creating Embryos Solely for Stem-Cell Research. *The Wall Street Journal*, January 18, 2002.
[8]Gouldner, Alvin. *The Dialectic of Ideology and Technology.* New York: Oxford University Press, 1976, p. 255.

penetrable as this specific structure is far removed from the academic situation of most bioethicists; however, it represents the perfect opportunity for bioethicists to introduce their concepts to help groups develop self-regulatory systems. Ethics needs to be made a part of the bureaucracy, through its ends or its means; however, the ideological focus of technocracies lies in technical development, and that alone will remain the end. Although the ideal situation is to make ethics as much an end as technology, it is unlikely that this shift will occur de novo. Therefore, efforts must be made to include ethical checkpoints. The product pipeline, research protocols, marketing, and organizational structures, and other systems must be analyzed to identify where to integrate analysis to weigh ethical conduct.

As new technologies arise in biotechnology, and as the field intersects with different types of technology (the Internet, etc.), the issues facing such advances will only increase. The voices that sound for and against these new applications will grow louder as possibilities arise that had, in the past, been grounded in science fiction rather than science itself. Whether they are special interest groups or regulatory and legal agencies, forces are assembling to engage a bioethical debate that affects the participants at a profound individual and cultural level. Despite the biotechnology industry's own concern regarding ethical conduct in their practices, the field of bioethics has taken the initiative to offer direction at the crossroads of industry, science, and society. The burden lies on the shoulders of industry, which must sieze the opportunity to reform itself while the choice remains in its hands. However, if bioethicists intend to maintain their position as mediators, they will have to step beyond their traditions and gain sympathy for the practices that define modern biological research. As such they will have greater appeal to industry, which will more clearly see the concrete benefits of improving society through responsibly delivered technology. For organizations unable to see merit in progressive action for its own sake, bioethicists will need to understand technocracies well enough to articulate the economic advantages linked to social and ecological concern;[9] for instance, balance sheets will positively reflect the anticipation of legislation and regulation. Through its integration into the functioning of scientific practice, ethical conduct may eventually take its place as simply a part of scientific practice.

---

[8]Elkington, John. *The Triple Bottom Line.* Vancouver, BC: New Society Publishers, 1998.

GLOSSARY

Adenovirus—a group of DNA-containing viruses that cause respiratory disease, including one form of the common cold. Adenoviruses can also be genetically modified and used in gene therapy to treat cystic fibrosis, cancer, and, potentially, other diseases.

Adult stem cells—an undifferentiated cell found in a differentiated tissue that can renew itself and differentiate to yield all the specialized cell types of the tissue from which it originated.

Adverse drug reaction—an unwanted, damaging physiological response to a drug; also known as a side effect.

Agbiotech—the division of biotechnology that focuses on agriculture; most commonly used in association with genetically modified plants.

Agriceutical—using biotechnology to alter plants so that the modified organism contains a pharmaceutical therapy, for instance, a vaccine.

Agrobacterium tumefaciens—a common soil bacterium that causes crown gall disease by transferring some of its DNA to the plant host. The transferred DNA (T-DNA) is stably integrated into the plant genome, where its expression leads to the synthesis of plant hormones and thus to the tumorous growth of the cells. This bacterium can be altered to transfer a specific gene into its host, genetically modifying the plant.

Amino acids—a group of 20 different kinds of small molecules that link together in long chains to form proteins. Often referred to as the "building blocks" of proteins.

Anonymization—removing personal identifiers from biological material while maintaining a translation key that allows whomever possesses it to link the information to the samples.

Antibiotic resistance—characterizes an organism that was once vulnerable to antibiotics but has since developed an immunity to the drug class or a particular drug.

Apolipoprotein—the major protein component of high-density lipoproteins (HDLs).

Autoimmune disorders—diseases caused by an immune response against the body's own tissues.

AZT (Zidovudiine)—a therapy for human immunodeficiency virus (HIV).

Bacillus thuringiensis—a soil bacterium that produces a series of proteins that repel common crop pests.

Balanced polymorphism—the maintenance of two different phenotypes of the same species in a population, usually because one phenotype confers a selective advantage, although specific environmental factors may cause it to be deleterious.

Big Pharma—large pharmaceutical companies with market capitalizations in the many billions of dollars.

Biodiversity—the variety of the world's organisms, including their genetic diversity and the assemblages they form.

Bioinformatics—the science of managing and analyzing biological data using advanced computing techniques.

Biolistic—a method of transferring DNA into a cell by "shooting" it into the cell.

Biomarker—a characteristic that is objectively measured and evaluated as an indicator of normal biological processes, pathogenic processes, or pharmacologic responses to a therapeutic intervention.

Biopiracy—the collecting and patenting of life forms formerly held in common and their exploitation for profit.

Blastocoel—the cavity in the blastula of the developing embryo.

Blastocyst—a preimplantation embryo consisting of 30–150 cells. The blastocyst consists of a sphere made up of an outer layer of cells (the trophectoderm), a fluid-filled cavity (the blastocoel), and a cluster of cells on the interior (the inner cell mass).

Blastula—an early stage in the development of the ovum consisting of a hollow sphere of cells enclosing a cavity known as the blastocoel.

Bovine spongiform encephalopathy (BSE, mad cow disease)—a transmissible, neurodegenerative, fatal brain disease of cattle.

Cell selection—selecting specific cells from a cell culture and transferring them to another culture to propagate a particular cell type.

Chromosome—one of the threadlike "packages" of genes and other DNA in the nucleus of a cell. Different kinds of organisms have different numbers of chromosomes. Humans have 23 pairs of chromosomes, 46 in all: 44 autosomes and two sex chromosomes. Each parent contributes one chromosome to each pair, so children get half of their chromosomes from their mothers and half from their fathers.

Cloning—1. The process of making copies of a specific piece of DNA, usually a gene. When geneticists speak of cloning, they do not mean the process of making genetically identical copies of an entire organism. 2. Replicating an organism based on its exact genetic code.

Coding—protecting the identity of a donor by assigning codes to a biological sample in lieu of personal identifiers.

Consequentialism—a school of ethical thought that assigns moral value to actions according to the nature of their consequences, thereby holding no particular act to be intrinsically right or wrong.

Contractarian ethics—a family of moral and political theories that maintains that good behavior is defined by foundational ideas agreed upon by those involved in social interaction.

Conversion—unlawfully turning or applying the personal goods of another to the use of the taker, or of some other person than the owner.

Creutzfeldt–Jakob disease—a fatal disease in humans, like BSE, that results in irreversible brain damage and, ultimately, death.

Crossbreeding—mixing different strains of crops (or other organisms) together based on desired traits exhibited by both in hopes that the offspring will contain only the desired traits.

Cultivar—cultivated variety of crop.

Cystic fibrosis—a genetic disease affecting the lungs, most cases of which are caused by a defect in a single gene.

Cytokines—proteins produced by white blood cells that act as chemical messengers between cells.

Decisional privacy—a type of privacy signifying the ability to make one's own decisions and to act on those decisions, free from governmental or other unwanted interference

Deontology—an ethical school of thought that holds that moral claims are as duties with an obligatory character and action in the moral domain is intrinsically right or wrong.

Diagnostic genetic testing—genetic testing to predict disease.

Dignatory harm—associated with deontology, this refers to a transgression that violates morality, whether there is a positive, negative, or neutral effect from an action.

DNA—deoxyribonucleic acid, the chemical inside the nucleus of a cell that carries the genetic instructions for making living organisms.

DNA bank—a repository of DNA samples.

DNA data bank—a repository of data derived from DNA samples.

DNA future diary—a term to describe the predictive power of information derived from DNA; sometimes called a probabilistic future diary because it describes an important part of a person's unique future and, as such, can affect and undermine an individual's view of his/her life's possibilities.

DNA sequencing—directly examining DNA to determine the order of nucleotides (base sequences) in a DNA or RNA molecule or the order of amino acids in a protein.

Ecosphere—areas inhabitable by organisms.

Embroyoid bodies—clumps of celluar structures that arise when embryonic stem cells are cultured; not part of normal development, they only arise in vitro.

Embryo—the developing organism from the time of fertilization until the end of the eighth week of gestation, when it becomes known as a fetus.

Embryonic stem cells—cells that have the ability to divide for indefinite periods in culture and to give rise to specialized cells.

Ethics advisory board (EAB)—a group of individuals in a company whose goals are to advise on ethical questions that face corporations.

Expression (gene and protein)—the process by which a gene's coded information is converted into the structures present and operating in the cell.

Forced transparency—a situation facing corporations by which their operations are made public by individuals who investigate their operations.

Gamete—mature male or female reproductive cell (sperm or ovum) with a haploid set of chromosomes (23 for humans).

Gene—a functional unit of heredity which is a segment of DNA located in a specific site on a chromosome. A gene directs the formation of an enzyme or other protein.

Gene flow—the potential for the DNA from a genetically engineered plant or animal to mix with wild type DNA, producing an undesirable genetic variant.

Gene gun—the tool used to execute the biolistic method of transforming plants; see biolistic.

Gene jumping—see gene flow.

Gene mapping—determination of the relative positions of genes on a DNA molecule (chromosome or plasmid) and of the distance, in linkage units or physical units, between them.

Gene therapy—an experimental procedure aimed at replacing, manipulating, or supplementing nonfunctional or malfunctioning genes with healthy genes; genetic engineering.

Genetic information—information that is revealed by analyzing inherited traits, either by family history/pedigrees or direct DNA analysis.

Genetically modified food—food produced by combining a desirable genetic trait from one species with the genome of another species.

Genetic reductionism—explaining a phenomenon solely in terms of genetics; in effect, emphasizing genetics over other types of explanations.

Genetics—the study of inheritance patterns of specific traits.

Genome—all the genetic material in the chromosomes of a particular organism; its size is generally given as its total number of base pairs.

Genomics—the study of genes and their function.

Genotype—the genetic identity of an individual that does not show as outward characteristics.

Golden Rice—a genetically modified rice that appears yellowish due to a gene that produces beta-carotene, a precursor of vitamin A.

Graft versus host disease—immune attack on the recipient prompted by cells from a donor.

Green Revolution—a movement to relieve hunger throughout the world based on nonrecombinant-DNA-based technologies that manipulate plant genetics.

Haplotype—a set of closely linked genetic markers present on one chromosome that tend to be inherited together (not easily separable by recombination).

Heterozygous—possessing two different forms of a particular gene, one inherited from each parent.

Homozygous—possessing two identical forms of a particular gene, one inherited from each parent.

Human Genome Project—an international research project to map each human gene and to completely sequence human DNA.

Hybrid—in agriculture, creating strains of plants from genetically dissimilar "parents."

Immunosuppression—the use of drugs or techniques to suppress or interfere with the body's immune system and its ability to fight infections or disease.

In vitro fertilization—an assisted reproduction technique in which fertilization is accomplished outside of the body.

Informational privacy—privacy related to access to information by third-party individuals.

Informed consent—consent to medical treatment by a patient, or to participation in a medical experiment by a subject, after achieving an understanding of the risks and benefits.

Intellectual property—the intangible value created by human creativity and invention; includes copyrights, trademarks, and patents.

Institutional review board (IRB)—a specially constituted review body established or designated by an entity to protect the welfare of human subjects recruited to participate in biomedical or behavioral research.

Kanamycin—an antibiotic.

Late embryonic abundant gene promoter—a DNA sequence that activates gene transcription for its adjacent gene, but only late in embryo development.

Linkage disequilibrium—alleles occurring together more often than can be accounted for by chance. Indicates that the two alleles are physically close on the DNA strand.

Lipid—a fatty molecule insoluble in blood.

Lipoprotein—a lipid surrounded by a protein; the protein makes the lipid soluble in blood.

Luddites—individuals who are opposed to technology; based on early nineteenth century English workmen who destroyed machinery in protest. The Luddites were the first organized movement to oppose the mechanized technology of the Industrial Revolution.

Mad cow disease—see BSE.

Methionine—an amino acid that is not synthesized by many mammals; thus, humans and animals must incorporate it through some edible source.

Morula—a solid mass of cells that results from the cleavage of an ovum.

Multipotent—capable of giving rise to a number of cell types specific to a particular germ layer.

Mutation—a permanent structural alteration in DNA. In most cases, DNA changes either have no effect or cause harm, but occasionally a mutation can improve an organism's chance of surviving and passing the beneficial change on to its descendants.

Nonresponders—individuals who do not respond to a drug therapy.

Off-patent drugs—drugs that are no longer protected by patents and thus exist in the public domain.

Nucleotide—a subunit of DNA or RNA consisting of a nitrogenous base (adenine, guanine, thymine, or cytosine in DNA; adenine, guanine, uracil, or cytosine in RNA), a phosphate molecule, and a sugar molecule (deoxyribose in DNA and ribose in RNA). Nucleotides are linked to form a DNA or RNA molecule.

Orphan diseases—rare diseases, which, by virtue of their limited population and profitability, do not receive great attention from drug development groups.

Orphan drugs—therapies developed for orphan diseases.

Personalized medicine—tailoring medical care to individuals by accounting for genetic and other types of variability.

Pharmacogenetics—the study of genetic variation underlying differential response to drugs.

Pharmacogenomics—the study of the interaction of an individual's genetic makeup and response to a drug.

Phenotype—the physical characteristics of an organism may or may not be genetic.

Phlebotomy—drawing blood.

Physical privacy—the right to privacy, protecting one from the invasion of one's own body.

Plasmid—autonomously replicating extrachromosomal circular DNA molecules, distinct from the normal bacterial genome and nonessential for cell survival under nonselective conditions. Some plasmids are capable of integrating into a host genome. A number of artificially constructed plasmids are used as cloning vectors.

Plant regeneration—growing an entire organism by nurturing cells from a single plant.

Pluripotency—capability of giving rise to most tissues of an organism.

Precautionary principle—a principle dictating that where there is threat of serious or irreversible environmental damage, lack of full scientific certainty should not be used as a reason for postponing measures to prevent environmental degradation. In the application of the precautionary principle, public and private decisions should be guided by careful evaluation to avoid, wherever practicable, serious or irreversible damage to the environment. An assessment of the risk-weighted consequences of various options.

Primordial germ cells—fetal cells that develop into the reproductive organs.

Promoter—the part of a gene that contains the information needed to turn the gene on or off. The process of transcription is initiated at the promoter.

Proprietary privacy—privacy relating to the right to control how one profits from one's own body.

Quality traits—traits engineered into plants that add benefits to the consumer—for instance, lower fat content or higher vitamin content.

Responders—individuals who react to a therapy in the expected manner.

Retrovirus—a type of virus that contains RNA as its genetic material. The RNA of the virus is translated into DNA, which inserts itself into an infected cell's own DNA. Retroviruses can cause many diseases, including some cancers and AIDS.

RNA (ribonucleic acid)—a chemical found in the nucleus and cytoplasm of cells; it plays an important role in protein synthesis and other chemical activities of the cell. The structure of RNA is similar to that of DNA. There are several classes of RNA molecules, including messenger RNA, transfer RNA, ribosomal RNA, and other small RNAs, each serving a different purpose.

Scorable marker—a marker gene added to a plant along with the gene conferring the desired trait. This marker gene reacts to a stimuli to indicate a successful transfer; for instance, the nutrient broth in which the cells are being cultured may cause the scorable marker gene's product to alter the color of the cells.

Selectable marker—a marker gene added to a plant along with the gene conferring the desired trait. This marker gene confers survival traits in the presence of some challenge that would normally kill an unmodified plant; thus, the cell culture's nutrient broth may contain an antibiotic, which would only allow cells to survive if they carried an inserted antiobiotic resistance gene.

Sickle cell anemia—a genetic disease in which the hemoglobin protein is mutated, giving red blood cells a twisted shape that may painfully block circulation. This often leads to medical crises and may cause an early death. The recessive disease occurs in children who have inherited the mutated gene from both their parents. Occurs mostly among people of African or Mediterranean origin.

Single nucleotide polymorphism (SNP)—DNA sequence variation that occurs when a single nucleotide (A, T, C, or G) in the genome sequence is altered.

Somaclonal variation (SCV)—heritable changes that result from in vitro procedures.

Somatic cell nuclear transfer (SCNT)—also known as cloning, this technique works by replacing the nucleus of an ovum with that of a somatic cell.

Stakeholder—any person with an interest in a corporation's products, operations, or overall functioning.

StarLink corn—a genetically modified corn that contains a gene to produce pesticidal proteins targeted at the corn borer.

Statistical genetics—a discipline that applies statistical methods to genetic data, both population and quantitative, to interpret DNA and protein sequence data, locate genes affecting quantitative traits and human diseases, and the significance of matching DNA profiles.

STEP—a tool to aid corporations account for the social, technical, environmental, and political influences that affect their operations.

Strong encryption—a method of data protection in which any translational keys are discarded after encryption, making it impossible to link the coded information to the data.

Telomere—the end of a chromosome. This specialized structure is involved in the replication and stability of linear DNA molecules.

Teosinte—a tall grass-like plant believed to be the ancestor of maize.

Terminator—a genetic modification technology that prohibits plants from producing seeds, thus copy-protecting the plant.

Therapeutic genetic testing—using genetic analysis to reveal an individual's ability to respond to a particular therapy.

Tissue bank—a repository of organic tissue.

Totipotent—having unlimited capability. Totipotent cells have the capacity to specialize into extraembryonic membranes and tissues, embryos, and all postembryonic tissues and organs.

Transcription—the synthesis of an RNA copy from a sequence of DNA (a gene); the first step in gene expression.

Transformation—introduction of an exogenous DNA molecule into a cell, causing it to acquire a new phenotype (trait).

Transgene—the exogenous gene inserted into an organism/cell to be incorporated into the host organism/cell's genome.

Transgenic—an organism that has been transformed with a foreign DNA sequence; an experimentally produced organism in which DNA has been artificially introduced and incorporated into the organism's germ line.

Translation—the process in which the genetic code carried by mRNA directs the synthesis of proteins from amino acids.

Transparency—revealing a corporation's internal operations and policies to the public.

Unlinking—removing personal identifiers and discarding any means of connecting these identifiers from their associated data/biological samples.

Vector (cloning vector)—DNA molecule originating from a virus, a plasmid, or the cell of a higher organism into which another DNA fragment of appropriate size can be integrated without loss of the vector's capacity for self-replication; vectors introduce foreign DNA into host cells, where the DNA can be reproduced in large quantities.

Wild type—the form of an organism that occurs most frequently in nature.

Xenotransplantion—transplanting a foreign tissue into another species.

Zygote—a fertilized egg. A diploid cell resulting after fertilization of an egg by a sperm cell.

SUGGESTED READING

Chapter 1. Introduction

1. Beauchamp, Tom L. and Childress, James F., *Principles of Biomedical Ethics,* fifth edition, Oxford University Press, New York, 2001.
2. Elkington, John, *Cannibals with Forks: The Triple Bottom Line of 21st Century Business,* New Society Publishers, Vancouver, BC, Canada, 1998.
3. Jonsen, Albert, *Birth of Bioethics,* Oxford University Press, New York, 1988.
4. Lewontin, Richard C., *Biology as Ideology,* HarperPerennial Library, New York, 1993.
5. Lewontin, Richard, C., *It Ain't Necessarily So: The Dream of the Human Genome and Other Illusions,* New York Review Books, New York, 2000.
6. Reilly, Philip R., *Abraham Lincoln's DNA and Other Adventures in Genetics,* Cold Spring Harbor Laboratory Press, Cold Spring Harbor, NY, 2000.
7. http://www.aslme.org/.
8. http://www.bio.org.
9. http://www.bioethics.gov.
10. http://www.nhgri.nih.gov.

Chapter 2. Genetically Modified Foods

1. Carson, Rachel, *Silent Spring,* Houghton Mifflin, Boston, 1962.
2. McHughen, Alan, *Pandora's Picnic Basket: The Potential and Hazards of Genetically Modified Foods,* Oxford University Press, New York, 2000.
3. Nelson, Gerald C., ed., *Genetically Modified Organisms in Agriculture: Economics and Politics,* Academic Press, New York, 2001.
4. Risler, Jane and Mellon, Margaret, *The Ecological Risks of Engineered Crops,* MIT Press, Cambridge, MA, 1996.

5. http://www.agbiotechnet.com.

6. http://www.aventis.com.

7. http://www.epa.gov.

8. http//www.fda.gov.

9. http://www.monsanto.com.

10. http://www.ncbe.reading.ac.uk.

11. http://www.ucsusa.org.

12. http://www.usda.gov.

Chapter 3. DNA Data Banking

1. Annas, George J., *Some Choice: Law, Medicine and the Market,* Oxford University Press, New York, 1998.

2. Berg, Jessica, Applebaum, Paul S., Parker, Lisa S., Lidz, and Charles W., *Informed Consent: Legal Theory and Clinical Practice,* Oxford University Press, New York, 2001.

3. Kevles, Daniel J. and Hood, Leroy, *The Code of Codes: Scientific and Social Issues in the Human Genome Project,* Harvard University Press, Cambridge, MA, 1993.

4. Rothstein, Mark A., ed., *Genetic Secrets: Protecting Privacy and Confidentiality in the Genetic Era,* Yale University Press, New Haven, CT, 1999.

5. Walters, Leroy and Palmer, Julie Gage, *The Ethics of Human Gene Therapy,* Oxford University Press, New York, 1996.

6. http://www.bioethics.gov.

7. http://www.dna.com.

8. The Common Rule. Office for Protection of Research Risks (now known as Office for Human Research Protections), *Common Rule, Protection of Human Subjects, Title 45, Code of Federal Regulations. Part 46,* Revised June 18, 1991, Reprinted April 2, 1995, Department of Health and Human Services.

9. The Declaration of Helsinki. World Medical Association, *Declaration of Helsinki: Recommendations Guiding Medical Doctors in Biomedical Research Involving Human Subjects,* revised in Edinburgh, Scotland, October, 2000. http://www.wma.net/e/policy/17-c_e.html.

10. The Nuremberg Code. *Trials of War Criminals before the Nuremberg Military Tribunals under Control Council Law No. 10:* Nuremberg, October 1946-1949, 2 vols., U.S. Government Printing Office, Washington, DC, 1949: See also: Annas, George J. and Grodin, Michael A., *The Nazi Doctors and the Nuremberg Code: Human Rights in Human Experimentation,* Oxford University Press, New York, 1992.

Chapter 4. Personalized Medicine

1. Baker, Diane L., Schuette, Jane L., and Uhlman, Wendy R., eds., *A Guide to Genetic Counseling*, Wiley, New York, 1998.

2. Broday, Baruch A, *Ethical Issues in Drug Testing, Approval and Pricing: The Clot-Dissolving Drugs*, Oxford University Press, New York, 1995.

3. Farmer, Paul, *Infections and Inequalities: The Modern Plagues*, University of California Press, Berkeley, CA, 2001.

4. Kessler, Seymour and Resta, Robert G., eds., *Psyche and Helix: Psychological Aspects of Genetic Counseling*, Wiley-Liss, New York, 2000.

5. Mundy, Alicia, *Dispensing with the Truth: The Victims, the Drug Companies and the Dramatic Story behind the Battle over Phen-Fen*, St. Martin's Press, New York, 2001.

6. Rothstein, Mark A., ed., *Genetic Secrets: Protecting Privacy and Confidentiality in the Genetic Era*, Yale University Press, New Haven, CT, 1999.

7. http://www.fda.gov.

8. http://www.genaissance.com.

9. http://www.ilgenetics.com.

Chapter 5. Stem Cells

1. Dworkin, Ronald, *Life's Dominion: An Argument about Abortion, Euthanasia, and Individual Freedom*, Vintage Books, New York, 1994.

2. Harris, John and Holm, Soren, *The Future of Human Reproduction: Ethics, Choice, and Regulation*. Issues in Biomedical Ethics, Oxford University Press, New York, 2000.

3. Holland, Suzanne, Lebacqz, Karen, and Zoloth, Laurie, eds., *The Human Embryonic Stem Cell Debate: Science, Ethics, and Public Policy*, MIT Press, Cambridge, MA, 2001.

4. Kass, Leon R. and Wilson, James Q., *The Ethics of Human Cloning*, AEI Press, Washington DC, 1998. http://www.aei.org/shop1/shops/1/4050-0.pdf.

5. Lauritzen, Paul, ed., *Cloning and the Future of Human Embryo Research*, Oxford University Press, New York, 2001. http://www.advancedcell.com/.

6. http://www.geron.com.

7. http://www.nih.gov/news/stemcell/primer.htm.

8. http://www.religioustolerance.org/abortion.htm.

INDEX